First-Order Methods in Large-Scale Semidefinite Optimization

First-Order Methods in Large-Scale Semidefinite Optimization

Diss. ETH No. 20409

First-Order Methods in Large-Scale Semidefinite Optimization

A dissertation submitted to
ETH ZURICH

for the degree of
DOCTOR OF SCIENCES

presented by
MICHAEL KLAUS BÜRGISSER
Dipl. Math., University of Zurich
born on September 24, 1982
citizen of Willisau (LU)

accepted on the recommendation of
Prof. Dr. Hans-Jakob Lüthi (examiner)
Prof. Dr. Yurii Nesterov (co-examiner)
Dr. Michel Baes (co-examiner)

2012

Bibliografische Information der Deutschen Nationalbibliothek
Die Deutsche Nationalbibliothek verzeichnet diese Publikation in der Deutschen
Nationalbibliografie; detaillierte bibliografische Daten sind im Internet über
http://dnb.d-nb.de abrufbar.
1. Aufl. - Göttingen: Cuvillier, 2012
Zugl.: (ETH) Zürich, Diss., 2012

978-3-95404-132-9

Acknowledgments

I wish to express my sincere gratitude to Prof. Hans-Jakob Lüthi, the examiner of this dissertation. I am deeply grateful that he gave me the opportunity and motivated me to carry out this thesis. I appreciate his continuous and encouraging support, the numerous helpful discussions, as well as the enjoyable and stimulating working atmosphere that he created. This last point includes also that Prof. Hans-Jakob Lüthi introduced me to several excellent researchers, which brings me to the next person I would like to thank.

I owe many thanks to Prof. Yurii Nesterov for being a co-examiner of this dissertation. In several meetings, I could immensely profit from his outstanding knowledge in Convex Optimization. I would like to thank him for these interesting scientific discussions, his helpful inputs, and his friendly support.

A special thank goes to Dr. Michel Baes, who supervised this thesis. Due to his great support, his deep understanding of Convex Optimization, and his unlimited enthusiasm for this project, he contributed to the success of this thesis in an essential way. In addition, I am very grateful to him for proofreading this dissertation carefully.

Numerous results of this dissertation have already been presented in [BB09, BB10, BB11, BBN11]. I show gratitude to my co-authors Prof. Arkadi Nemirovski and Dr. Michel Baes for these fruitful collaborations. I am particularly indebted to Prof. Arkadi Nemirovski for leading the joint research project that resulted in the paper [BBN11]. Without any doubt, he was research-wise the driving force in this joint project.

I further owe many thanks to the Swiss National Science Foundation for partially funding the research that led to this dissertation.

Special thanks go to my current and former colleagues at the Institute for Operations Research at ETH. It has not only been a pleasure to work with them, but I also enjoyed the many joint cultural, social, and sportive activities. As representative for all these activities, I would like to thank Annina, Dennis, Jan, Kathrin, Lorenz, Marco, Martin, Michael, Rico, Sabrina, Sarah, and Silvia for the exciting floorball matches on Monday evenings.

I am very grateful to my parents, my sister, and my brother for their support and their encouragement.
Finally, I would like to thank my friends who have always been there for me.

Michael Bürgisser, April 2012

Abstract

S EMIDEFINITE Optimization has attracted the attention of many researchers over the last twenty years. It has nowadays a huge variety of applications in such different fields as Control, Structural Design, Statistics, or in the relaxation of hard combinatorial problems. In this thesis, we focus on the practical tractability of large-scale semidefinite optimization problems. From a theoretical point of view, these problems can be solved by polynomial-time Interior-Point methods approximately. The complexity estimate of Interior-Point methods grows logarithmically in the inverse of the solution accuracy, but with the order 3.5 in both the matrix size and the number of constraints. The later property prohibits the resolution of large-scale problems in practice.

In this thesis, we present new approaches based on advanced First-Order methods such as Smoothing Techniques and Mirror-Prox algorithms for solving structured large-scale semidefinite optimization problems up to a moderate accuracy. These methods require a very specific problem format. However, generic semidefinite optimization problems do not comply with these requirements. In a preliminary step, we recast slightly structured semidefinite optimization problems in an alternative form to which these methods are applicable, namely as matrix saddle-point problems. The final methods have a complexity result that depends linearly in both the number of constraints and the inverse of the target accuracy.

Smoothing Techniques constitute a two-stage procedure: we derive a smooth approximation of the objective function at first and apply an optimal First-Order method to the adapted problem afterwards. We present a refined version of this optimal First-Order method in this thesis. The worst-case complexity result for this modified scheme is of the same order as for the original method. However, numerical results show that this alternative scheme needs much less iterations than its original counterpart to find an approximate solution in practice. Using this refined version of the optimal First-Order method in Smoothing Techniques, we are able to solve randomly generated matrix saddle-point problems involving a hundred matrices of size $12'800 \times 12'800$ up to an absolute accuracy of 0.0012 in about four hours.

Smoothing Techniques and Mirror-Prox methods require the computation of one or two matrix exponentials at every iteration when applied to the matrix saddle-point problems obtained from the above transformation step. Using standard techniques, the efficiency estimate for the exponentiation of a symmetric matrix grows cubically in the size of the matrix. Clearly, this operation limits the class of problems that can be solved by Smoothing Techniques and Mirror-Prox methods in practice. We present a randomized Mirror-Prox method where we replace the exact matrix exponential by a stochastic approximation. This randomized method outperforms all its competitors with respect to the theoretical complexity estimate on a significant class of large-scale matrix saddle-point problems. Furthermore, we show numerical results where the randomized method needs only about 58% of the CPU time of the deterministic counterpart for solving approximately randomly generated matrix saddle-point problems with a hundred matrices of size 800×800.

As a side result of this thesis, we show that the Hedge algorithm – a method that is heavily used in Theoretical Computer Science – can be interpreted as a Dual Averaging scheme. The embedding of the Hedge algorithm in the framework of Dual Averaging schemes allows us to derive three new versions of this algorithm. The efficiency guarantees of these modified Hedge algorithms are at least as good as, sometimes even better than, the complexity estimates of the original method. We present numerical experiments where the refined methods significantly outperform their vanilla counterpart.

Zusammenfassung

DIE Semidefinite Optimierung hat in den letzten 20 Jahren das Interesse unzähliger ForscherInnen auf sich gezogen und weist heutzutage eine Vielzahl von Anwendungen in den verschiedensten Gebieten wie Regelungstechnik, Baukonstruktion oder Statistik auf. Ausserdem werden schwierige kombinatorische Probleme häufig durch semidefinite Hilfsprobleme approximiert. Der Fokus dieser Arbeit liegt auf dem möglichst raschen Lösen von grossen semidefiniten Optimierungsproblemen. Theoretisch können diese Probleme approximativ mit der Inneren-Punkte Methode gelöst werden, da dieser Algorithmus eine polynomielle Laufzeit hat. Die schlimmst mögliche theoretische Laufzeit dieser Methode wächst logarithmisch mit dem Inversen der Fehlertoleranz und mit der Potenz 3.5 bezüglich der Grösse der Entscheidungsmatrix und der Anzahl der Nebenbedingungen. In der Praxis ist tatsächlich eine relativ schnelle, obwohl polynomielle, Zunahme der Rechenzeit bezüglich der Matrixgrösse und der Anzahl der Nebenbedingungen beobachtbar. Dadurch ist die Innere-Punkte Methode ungeeignet für die numerische Lösung grosser semidefiniter Optimierungsprobleme.

In dieser Arbeit werden neue Methoden vorgestellt, die leicht strukturierte, grosse semidefinite Optimierungsprobleme bis auf einen moderaten Approximationsfehler lösen können. Diese Methoden basieren auf fortgeschrittenen Subgradienten Algorithmen wie Smoothing Techniques oder Mirror-Prox Methoden, die aber nur auf Probleme mit einer ganz speziellen Struktur angewendet werden können. Jedoch weicht die Form der semidefiniten Optimierungsprobleme, die in dieser Dissertation betrachtet werden, von der vorausgesetzten Struktur ab. In einem ersten Schritt werden die Ausgangsprobleme daher in eine passende Form umgewandelt und als Sattelpunktprobleme formuliert, auf welche die fortgeschrittenen Subgradienten Methoden angewendet werden können. Die theoretischen Laufzeiten der so erhaltenen Methoden wachsen linear in der Anzahl der Nebenbedingungen und im Inversen der Fehlergenauigkeit.

Smoothing Techniques bestehen aus zwei Schritten. Zuerst wird eine ausreichend differenzierbare Approximation der Zielfunktion hergeleitet. Danach wendet man auf das neue Problem eine optimale Subgradienten Methode an. In dieser Dissertation wird eine verfeinerte Version dieser optimalen Subgradienten Methode hergeleitet. Die theoretische Laufzeit des adaptierten Algorithmus ist im ungünstigen Fall von der gleichen Grössenordnung wie die Laufzeit der ursprünglichen Methode. In der Praxis ist allerdings eine signifikante Beschleunigung beobachtbar, da der angepasste Algorithmus bedeutend weniger Iterationen als die ursprünglichen Methode braucht. Durch die Integration dieser abgeänderten Methode in Smoothing Techniques sind zufällig erzeugte Instanzen von Sattelpunktproblemen mit Matrizen der Grösse $12'800 \times 12'800$ bis auf einen absoluten Fehler von 0.0012 in ungefähr vier Stunden lösbar.

Wenn die fortgeschrittenen Subgradienten Methoden auf die Sattelpunktprobleme, die durch die Transformation der ursprünglichen semidefiniten Optimierungsprobleme entstehen, angewendet werden, muss pro Iteration mindestens ein Matrixexponential berechnet werden. Die Rechenzeit für eine solche Operation wächst kubisch mit der Grösse der symmetrischen Matrix, vorausgesetzt, dass das Exponential über die Diagonalisierung der symmetrischen Matrix berechnet wird. Diese Operation wird in der Praxis offensichtlich zum kritischen Faktor, falls Smoothing Techniques oder Mirror-Prox Methoden zum Lösen von grossen Instanzen der betrachteten Sattelpunktprobleme eingesetzt werden sollen. In dieser Dissertation wird eine randomisierte Mirror-Prox Methode präsentiert, bei der das Matrixexponential durch eine stochastisch erzeugte Approximation ersetzt wird. Dieser randomisierte Algorithmus unterbietet alle existierenden Methoden bezüglich der theoretischen Laufzeit auf einer signifikanten Subklasse von grossen Sattelpunktproblemen mit Matrizen. Diese theoretischen Resultate werden durch Beobachtungen in der Praxis belegt: Die randomisierte Methode benötigt etwa 58% der Rechenzeit des deterministischen Pendants für das approximative Lösen von zufällig erzeugten Instanzen mit Matrizen der Grösse 800×800.

Als Nebenresultat dieser Dissertation wird der Hedge Algorithmus – eine Methode, die in der Theoretischen Informatik weit verbreitet ist – im Kontext von Dual Averaging Methoden eingebettet und diskutiert. Basierend auf dieser Interpretation werden drei neue Versionen des ursprünglichen Hedge Algorithmus hergeleitet. Die theoretischen Komplexitätsresultate der neu entwickelten Algorithmen sind gleich gut oder gar besser als die Effizienzgarantien der bisherigen Methode. Es werden numerische Resultate präsentiert, bei denen diese neuen Methoden zu deutlich besseren Resultaten führen als der ursprüngliche Hedge Algorithmus.

Contents

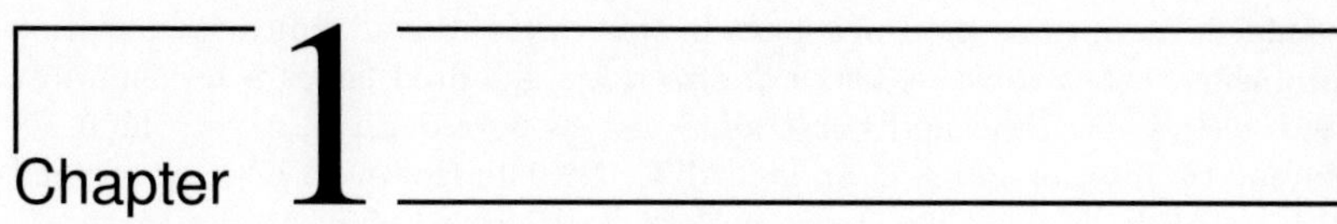

Chapter 1

Introduction

1.1 Motivation

Semidefinite Optimization is generally appreciated as the most exciting innovation in Convex Programming in the 1990's; see for instance [Fre99]. It constitutes a generalization of Linear Programming, where the cone of nonnegative vectors is replaced by the cone of positive semidefinite matrices. That is, we are supposed to minimize a linear function over a set of positive semidefinite matrices, where this set of matrices is described by a collection of linear inequalities.

Nowadays, an impressive amount of real-life optimization problems, from nearly all fields of engineering, can be represented or approximated by semidefinite optimization problems. For instance, Semidefinite Optimization is used in Control [BGFB94], in Structural Design [BTN97], or in Statistics [BV04], only to name a few. Probably the most famous applications are semidefinite relaxations of hard combinatorial problems; see for instance [GW95, NRT99].

Many modern real-life optimization problems – including up-to-date semidefinite optimization problems – are of very large-scale. Generally speaking, the emergence of large-scale optimization problems has two reasons. One the one hand, it has become relatively easy and cheap to collect and store huge amounts of data over the last few years. For instance, we may mention here web-based social platforms, customer bonus cards, or surveillance cameras, where massive amounts of data are collected every day. On the other hand, the real-life systems that we study and try to model as optimization problems are getting more and more complex. The relatively new field of Systems Biology serves as a prime example here. In Systems Biology, researcher try to analyze, reconstruct, and understand highly complex biological systems using – among others – optimization tools; see for instance [HNTW09, PRP04, TS08].

Motivated by these facts, we study the practical tractability of large-scale semidefinite optimization problems in this thesis. Semidefinite optimization problems that involve matrices of size a few hundred times a few hundred and with a few thousand constraints can be solved up to a very high accuracy by Interior-Point methods [Ali95, NN93]. However, if we go beyond this problem size, it takes Interior-Point methods too long to derive an approximate solution in practice. More formally, Interior-Point methods have a theoretical worst-case running time that is of the order

$$\mathcal{O}\left(\sqrt{m+n}\left[mn^3 + m^2n^2 + m^3\right]\ln\left[(m+n)/\epsilon\right]\right),$$

where m denotes the number of constraints, n the matrix size, and ϵ the target accuracy; see Chapter 6 for a review. The logarithmic dependence on ϵ makes Interior-Point methods a tailored tool for finding highly accurate solutions, whereas the fast – although polynomial – growth in m and n limits the size of problems that can be handled in practice. In order to comply with the modern trends in Semidefinite Optimization, that is, in order to be able to solve large-scale semidefinite optimization problems in practice, the following question arises immediately:

"Assume that we tolerate a moderate complexity increase with respect to the solution accuracy ϵ, say from $\ln[1/\epsilon]$ to $1/\epsilon$, or even to $1/\epsilon^2$. Does there exist an algorithm for solving semidefinite optimization problems with a running time that is below the complexity estimate of Interior-Point methods with respect to the matrix size and the number of constraints?"

This is the opening question of this thesis. Note that a moderate solution accuracy is usually not a barrier in practice, as two or three accuracy digits are typically sufficient for practical applications.

In 2007, Arora and Kale [AK07] introduced an alternative approach for solving slightly structured large-scale semidefinite optimization problems. They perform a Binary Search over the objective function values. At every iteration of this search, they are supposed to answer a feasibility question. The answer to this question is derived by a Matrix Multiplicative Weights Update method. In total, they need to perform, roughly speaking, $\mathcal{O}\left(\ln[1/\epsilon]/\epsilon^2\right)$ iterations of this method in order to find a solution with approximation error $\epsilon > 0$. By "roughly speaking", we refer to the fact that the complexity result depends also on other problem parameters such as the scaling of the problem. At every iteration of this method, the exponentiation of a symmetric matrix and some other computations not exceeding the cost of $\mathcal{O}(mn^2)$ arithmetic operations are supposed to be carried out. Provided that we compute the exponential through an eigendecomposition of the symmetric matrix, we end up with a method that requires in total, roughly speaking,

$$\mathcal{O}\left([n^3 + mn^2]\ln[1/\epsilon]/\epsilon^2\right)$$

arithmetic operations to compute an approximate solution. In case of sparse matrices, the term mn^2 in the above complexity result can be reduced in accordance to the sparsity pattern of the matrices. Arora and Kale [AK07] also discuss some strategies to replace the exact value of the matrix exponential by a random approximation in their Matrix Multiplicative Weights Update method. For instance, they present a random approximation that is based on the Johnson-Lindenstrauss Lemma [JL84] and on an appropriate truncation of the matrix exponential Taylor series. Using this randomization procedure, we end up with an algorithm whose complexity grows with the order of $\mathcal{O}(\ln[1/\epsilon]/\epsilon^5)$ with respect to solution accuracy $\epsilon > 0$. Because of this fast growth in ϵ, we do not elaborate more on this randomization of their method in this thesis.

The Matrix Multiplicative Weights Update method, which was introduced simultaneously by Arora and Kale [AK07] and by Warmuth et al. [TRW05, WK06], can be seen as a generalization of Multiplicative Weights Update methods to matrices; see [AHK05] for a survey of these methods. The basic concept of Multiplicative Weights Update methods plays a crucial role in the development of algorithms in Machine Learning and in Data Mining, or, more generally, in Computer Science. For instance, AdaBoost [FS97] – one of the top ten Data Mining algorithms (see [WKR$^+$07] for the complete list) – is based on the Hedge algorithm [FS97], which follows the same basic construction as Multiplicative Weights Update methods. Interestingly enough, the Hedge algorithm and (Matrix) Multiplicative Weights Update methods have the same analytical complexity as Dual Averaging schemes [Nes09]: the iteration count of all these methods grows with the order $\mathcal{O}(1/\epsilon^2)$. This observation gives rise to the conjecture that the Hedge algorithm and (Matrix) Multiplicative Weights Update methods are particular instances of Dual Averaging schemes.

Let us now get back to Semidefinite Optimization. When we compare the complexity results of Interior-Point methods and Arora and Kale's scheme for solving slightly structured semidefinite optimization problems, we make the following two observations. On the one hand, the complexity of Arora and Kale's method grows only linearly in the number of constraints. This is in sharp contrast to Interior-Point methods, where the complexity result grows with the power 3.5 with respect to m. On the other hand, there is a tremendous gap (namely, a factor of $1/\epsilon^2$) in the worst-case complexity bounds of

Interior-Point methods and Arora and Kale's scheme. These observations result in the following adaption of the opening question:

"Does there exist a method for solving slightly structured semidefinite optimization problems whose iteration count grows, roughly speaking, with the order $\mathcal{O}(1/\epsilon)$ and whose iteration cost is given by $\mathcal{O}(n^3 + mn^2)$?"

Smoothing Techniques [Nes05] and Mirror-Prox methods [Nem04a] were introduced recently. These powerful methods can be applied to non-differentiable convex problems that have a very specific structure, namely to a huge variety of matrix saddle-point problems. The matrix saddle-point problems that we consider in this thesis correspond all to the problem of minimizing the maximal eigenvalue of convex combinations of symmetric matrices. When we speak of matrix saddle-point problems, we henceforth automatically refer to the maximal eigenvalue minimization problem. Interestingly enough, the iteration count of Smoothing Techniques and Mirror-Prox methods grows with the desired order $\mathcal{O}(1/\epsilon)$, where ϵ denotes the solution accuracy. However, these powerful methods are not directly applicable to generic semidefinite optimization problems, as these problems do not satisfy the structural requirements of the algorithms. Nevertheless, this approach seems to be very promising, because Chudak and Eleutério [CE05] successfully applied these methods to large-scale linear optimization problems with up to millions of variables and constraints. As linear programs constitute a particular subclass of semidefinite optimization problems, it seems to be very natural to extend their approach to the more general class.

In the situation where the decision variables are matrices, Smoothing Techniques and Mirror-Prox methods suffer the same computational bottleneck as the Matrix Multiplicative Weights Update algorithm: they need to determine the exponential of a symmetric $(n \times n)$-matrix at every iteration; see [Nes07] and [Nem04a] for the details. There exists plenty of different ways to determine or to approximate these exponentials; see [ML03] for the classical survey on this topic. Standardly, this exponentiation is performed through an eigendecomposition of the symmetric matrix, requiring $\mathcal{O}(n^3)$ arithmetic operations and, consequently, hampering the resolution of problems with huge matrices. In order to extend the class of matrix saddle-point problems that can be successfully handled by Smoothing Techniques or Mirror-Prox methods in practice, we need to find strategies that allow us to replace this matrix exponential by an approximation that can be computed faster.

1.2 Goals of the thesis

Let us list the goals of this thesis.

A. Apply Smoothing Techniques and Mirror-Prox methods to semidefinite optimization problems

In order to close the dramatic gap between the complexity results of Interior-Point methods and Arora and Kale's scheme [AK07] with respect to the solution accuracy, we want to apply Smoothing Techniques and Mirror-Prox methods to slightly structured semidefinite optimization problems. However, these problems do not match the structural requirements of Smoothing Techniques and Mirror-Prox methods. In a preliminary step, we therefore need to find an appropriate reformulation of semidefinite optimization problems, that is, we are supposed to recast them as matrix saddle-point problems.

B. Reduce the number of iterations of Smoothing Techniques in practice

Smoothing Techniques are a two-stage procedure. In a first step, an appropriate smooth approximation of the objective function is built. This construction exploits the very specific form by which non-differentiability enters the problem. In a second step, we apply an optimal First-Order method [Nes04, Nes05] to the smooth auxiliary problem. At every iteration of this optimal First-Order method, the Lipschitz constant of the gradient of the smoothed objective function is used to determine the next iterate. Clearly, this constant is a global parameter of the problem and might be very pessimistic for the local environment of the algorithm's current iterate. We want to derive a strategy that allows us to replace this global constant by local estimates in the optimal First-order method [Nes04, Nes05].

C. Reduce the iteration cost in Mirror-Prox methods

As pointed out above, both Smoothing Techniques and Mirror-Prox methods require the computation of a matrix exponential at every iteration when applied to matrix saddle-point problems. When we solve problems with huge matrices, this operation becomes critical with respect to the running time of the method. We want to overcome this difficulty by replacing the exact value of the matrix exponential in Mirror-Prox methods by a randomized approximation that can be computed faster. We perform this discussion for Mirror-Prox methods, as this topic was studied in a joint project with Arkadi Nemirovski, the designer of Mirror-Prox methods.

D. Interpret the Hedge algorithm as a Dual Averaging scheme

As shown in [CBL06], the Hedge algorithm can be seen as a particular instance of Mirror-Descent methods [NY83], which are a subclass of Dual Averaging schemes. In this thesis, we want to study this interpretation, point out its severe inconsistencies, and give a complete and consistent discussion of the Hedge algorithm in the context of Dual Averaging schemes. In particular, given the knowledge we gain from this new perspective on the Hedge algorithm, we hope to define alternative versions of this scheme, which perform even more successfully in theory and in practice than the vanilla method.

1.3 Structure of the thesis

This thesis is split in three parts.

Part I: Analytical complexity of solution methods in Convex Optimization

In Part I of this thesis, we lay the theoretical and methodical foundation of this thesis. We introduce general optimization problems, review all properties of both the objective function and the feasible set which are relevant for this thesis, and study the computational tractability of optimization problems in Chapter 2. In Chapter 3, we discuss some Black-Box Optimization methods, namely Dual Averaging schemes, Primal-Dual Subgradient methods, Mirror-Descent algorithms, and optimal First-Order methods. In particular, we present a refinement of optimal First-Order methods which complies with Goal B. We conclude Part I by reviewing Smoothing Techniques, Mirror-Prox schemes, and Interior-Point methods in Chapter 4.

Part II: A new perspective on the Hedge algorithm

Part II consists only of Chapter 5 and addresses exclusively Goal D. In this chapter, we recast the Hedge algorithm in the context of Dual Averaging schemes and derive three new versions of this method, which have theoretical convergence guarantees that are better or at least as good as the convergence result for the vanilla scheme. Numerical results show that all these modified methods perform better than their original counterpart in practice.

Part III: Approximately solving large-scale semidefinite optimization problems

Part III represents the core of this thesis. In Chapter 6, we give an introduction to large-scale Semidefinite Optimization and derive the full complexity result of Interior-Point methods for semidefinite optimization problems. We recast the Matrix Multiplicative Weights Update method in the context of Dual Averaging schemes and discuss the approach of Arora and Kale [AK07] for solving slightly structured semidefinite optimization problems in Chapter 7. The first part of this chapter can be seen as a generalization of some of the results from Chapter 5 to a matrix setting. In Chapter 8, we achieve Goal A by reformulating slightly structured semidefinite optimization problems in a form to which we can apply not only Primal-Dual Subgradients methods and Mirror-Descent schemes, but also advanced tools such as Smoothing Techniques and Mirror-Prox methods. That is, we recast these problems as matrix saddle-point problems. In the same chapter, we discuss the complexity of Mirror-Descent schemes when applied to these problems. We particularize Smoothing Techniques for matrix saddle-point problems in Chapter 9. In particular, we obtain a procedure that can be used to solve slightly structured semidefinite optimization problems. Importantly, the complexity estimate of this procedure is of the form $\mathcal{O}(1/\epsilon)$ with respect to the solution accuracy $\epsilon > 0$. In Chapter 10, we study the application of Mirror-Prox methods to matrix saddle-point methods and discuss a randomized computation of matrix exponential approximations. We conclude this part by showing some numerical results in Chapter 11.

The conclusions and an outlook are presented in Chapter 12. In the Appendix, we give a short introduction to regular norms and collect some technical proofs.

Part I

Analytical complexity of solution methods in Convex Optimization

Convex Optimization and computational tractability

"In our opinion, the main fact, which should be known to any person deal-ing with optimization models, is that in general optimization problems are unsolvable."

(Yurii Nesterov, 2004; see Page xv in [Nes04])

EVERY human being has to make several decisions every day. Some of them are of harmless consequences, others may seriously affect the de-cision maker and her environment in a positive or negative way, maybe even for a long time. Given a set of alternatives, called the **feasible set**, any rea-sonable decision maker wants to find an alternative with the most favorable utility for her and/or her environment. The utility is expressed in the form of an **objective function**, that is, a function that assigns a value to any alternative from the feasible set. The decision maker's target is now to find an element from the feasible set that maximizes the objective function, that is, the utility, over this set.

We give a mathematical reformulation of this decision problem in Section 2.1. Recasting a real-life decision problem in such a mathematical form is often a time-consuming, intensive, but also highly interesting process that requires a lot of creativity, experience in system design, and reflection on the current real-life situation. However, not any reformulation of the real-world problem as a mathematical optimization problem is computationally tractable. Sys-tem design (or mathematical modeling) is typically a trade-off between com-putational tractability and accurate reproduction of the real-world situation.

Loosely speaking, computational intractability means that the computation time required to find such an element with the most favorable utility grows prohibitively fast with respect to certain problem parameters. A proper introduction of the notion of computational (in)tractability is given in Section 2.2.

In Section 2.3, we study convexity and other properties of both the feasible set and the objective function. If both the feasible set and the objective function are convex, we say that an optimization problem is **convex**. Convex optimization problems form a particularly favorable family of optimization problems. As we show in Section 2.4, convex optimization problems are computationally tractable under very mild assumptions. This in sharp contrast to the computational tractability of general optimization problems, as no efficient method is known for solving them.

Finally, in Section 2.5, we introduce the notion of Bregman distances. Bregman distances are a tool to represent the geometric structure of the feasible set more adequately than, for instance, the standard Euclidean metric. This tool plays an important role in the subsequent chapters, where we study algorithms that can be applied to convex optimization problems.

Contributions and relevant literature: This chapter is a review of existing basic definitions and results.   We mainly use the references [BTN01, HUL93, Nem04a, Nes04, Nes07, Nes09, Roc70]. Parts of this review chapter are taken from [BBN11].

2.1 Mathematical formulation of an optimization problem

In this section, we give a mathematical reformulation of the real-life decision problem that we sketched above.

The alternatives in the above real-life decision problem are modeled as real vectors of length n, implying that the set of alternatives corresponds to a subset of $\mathbb{R}^n$. We refer to this set as **feasible set** and denote it by $Q \subset \mathbb{R}^n$. The elements of Q are called feasible points. The negative utility of a feasible point $x \in Q$ is expressed (or modeled) as the function value $f(x)$, where $f : \mathbb{R}^n \to \mathbb{R} \cup \{+\infty\}$ may attain any real number or $+\infty$. We refer to f as the **objective function**. Solving the real-life decision problem corresponds now to minimize[1] the objective function over the feasible set, which can be formulated as the optimization problem:

$$f^* := \inf_{x \in Q} f(x). \tag{2.1}$$

---

[1] Note that minimizing the negative utility corresponds to maximizing the utility.

Throughout the thesis, we assume that Problem (2.1) is solvable, that is, there exists a point $x^* \in Q$ such that $f^* = f(x^*) < +\infty$.[2] In particular, the "inf"-expression in (2.1) can be replaced by a "min". Any point $x^* \in Q$ that satisfies $f^* = f(x^*)$ is called an **optimal solution** to (2.1), and we refer to f^* as the **optimal value** to (2.1). Identifying an optimal decision to the mathematical formulation of the real-life problem is now equivalent to finding a feasible point x^* that satisfies $f^* = f(x^*)$.

Example 2.1 (Linear Optimization) *A linear optimization problem, or Linear Program, is of the form*

$$\min_{x} \left\{ c^T x : Ax \leq b, \; x \in \mathbb{R}^n_{\geq 0} \right\}, \tag{2.2}$$

where $A \in \mathbb{R}^{m \times n}$, $b \in \mathbb{R}^m$, $c \in \mathbb{R}^n$, and $\mathbb{R}_{\geq 0} := \{x \in \mathbb{R}^n : x \geq 0\}$. It is called linear, as both the objective function and the feasible set are described solely by linear functions.

2.2 What does computational tractability mean?

In order to find an optimal solution to Problem (2.1), we would like (and we typically need) to take advantage of the immense computation power of modern computers. In this section, we discuss the details of the use of computers for finding such a solution to (2.1), that is, we specify the computer model we want to use, the input to the computer, the code that we run, and the output of the computer. Importantly, this discussion will enable us to specify what computational tractability of optimization problems means. We follow closely Sections 5.1.2, 5.4.2, and 6.1 in the book of Ben-Tal and Nemirovski [BTN01].

At first, we need to clarify what we expect from the computer. In this thesis, we use the **real arithmetic computer** model, which corresponds to an idealized variant of the common computer. In this model, the computer can store countably many real numbers. In addition, the computer is able to perform exactly the elementary operations, which are addition, subtraction, multiplication, division, evaluation of standard functions such as $\exp(\cdot)$, and comparisons. Although these requirements significantly differ from the properties of a real-world computer, this model is commonly used in Convex Optimization. The main reason for its popularity is that we can avoid annoying formal details by choosing this model.

---

[2] For the problems that we study later on, we can explicitly guarantee the existence of an optimal solution. To simplify our exposition, we thus assume the existence of an optimal solution in this theory review and do not elaborate on infeasible or unbounded problems.

As next, we discuss the **encoding** of Problem (2.1). We never study the performance of a solution method with respect to a single instance of Problem (2.1), but rather for a family of problem instances. This family $\mathcal{P}$ is defined by the common structure of all problem instances $p \in \mathcal{P}$. Any instance $p \in \mathcal{P}$ is encoded by a finite-dimensional real data vector $\text{Data}(p)$. This vector comprises the dimension parameters of the problem and the coefficients of the generic analytic expressions for $f(x)$ and Q. In order to make the encoding clear, we give an example below. We define the **size** of the instance $p \in \mathcal{P}$ as the dimension of the data vector, that is:

$$\text{Size}(p) := \dim(\text{Data}(p)).$$

Example 2.2 (Ex. 2.1 continued; see Section 5.1.2 in [BTN01])
Any instance p of the family $\mathcal{LP}$ of Linear Optimization Problems (2.2) is encoded as follows:

$$\text{Data}(p) = [n, m, a_{11}, \ldots, a_{1n}, a_{21}, \ldots, a_{m1}, \ldots, a_{mn}, b_1, \ldots, b_m, c_1, \ldots, c_n],$$

where $A = (a_{ij})_{i,j=1}^{m,n}$. We obtain:

$$\text{Size}(p) = 2 + mn + m + n.$$

From the previous section, we know that the resolution of Problem (2.1) is identical to finding an optimal solution. However, this target can typically not be achieved. Generally, it is impossible to compute an exact real optimal solution to an arbitrary instance of (2.1) using finitely many elementary operations. Fortunately, this is not a practical restriction at all, as "approximate" solutions are sufficient for all practical purposes. No decision maker is able to implement a solution with an infinite number of digits in practice. Let us now formalize what we mean by "approximate" solutions.

Let $\mathcal{P}$ be a family of Optimization Problems (2.1), $p \in \mathcal{P}$, and $x \in \mathbb{R}^n$. We assume that this family is equipped with an **infeasibility measure** $\text{Infeas}_{\mathcal{P}}(x, p)$. This measure complies with the requirements:

1. $\text{Infeas}_{\mathcal{P}}(x, p) \geq 0$.

2. $\text{Infeas}_{\mathcal{P}}(x, p) = 0$ if and only if x is feasible for p.

3. For any $\lambda \in [0, 1]$ and any $y \in \mathbb{R}^n$:

$$\text{Infeas}_{\mathcal{P}}(\lambda x + (1 - \lambda)y, p) \leq \lambda \text{Infeas}_{\mathcal{P}}(x, p) + (1 - \lambda)\text{Infeas}_{\mathcal{P}}(y, p).$$

Example 2.3 (Ex. 2.1 continued; see Section 5.1.2 in [BTN01])
For the family $\mathcal{LP}$ of linear optimization problems, we can take:

$$\text{Infeas}_{\mathcal{LP}}(x, p) = \max \left[0, \max_{1 \leq j \leq m} \{(Ax)_j - b_j\}, \max_{1 \leq i \leq n} \{-x_i\} \right],$$

where $x \in \mathbb{R}^n$ and $p \in \mathcal{LP}$.

We are able to define approximate solutions properly.

Definition 2.1 (ϵ-solution) *Let $\epsilon > 0$. We call $\hat{x} \in \mathbb{R}^n$ an ϵ-solution to (2.1) if it satisfies at the same time $\mathrm{Infeas}_{\mathcal{P}}(\hat{x}, p) \leq \epsilon$ and $f(\hat{x}) - f^* \leq \epsilon$.*

A **feasible ϵ-solution** to (2.1) is consequently a point $\hat{x} \in Q$ that complies with the condition $f(\hat{x}) - f^* \leq \epsilon$. Alternatively, we may define approximate solutions in relative scale:

Definition 2.2 (Feasible ϵ-solution in relative scale) *Let $\epsilon \in (0, 1)$ and $f^* > 0$. We refer to $\hat{x} \in Q$ as a feasible ϵ-solution in relative scale to (2.1) if $f(\hat{x}) - f^* \leq \epsilon f^*$, that is, if $f(\hat{x}) \leq (1 + \epsilon) f^*$.*

Note that we display here the strong version for the definition of approximate solutions in relative scale. There exists also a weak version of this notion, where we replace $f(\hat{x}) - f^* \leq \epsilon f^*$ in the above definition by the condition $f(\hat{x}) - f^* \leq \epsilon f(\hat{x})$, that is, by $(1 - \epsilon) f(\hat{x}) \leq f^*$. Clearly, any point $x \in Q$ that satisfies $f(x) - f^* \leq \epsilon f^*$ fulfills also the requirement $f(x) - f^* \leq \epsilon f(x)$, as $f(x) \geq f^*$ by the definition of f^*. However, the opposite statement is not true, which explains the attributes "weak" and "strong".

In both the absolute and the relative scale, the ultimate goal is to control the **optimality gap**: given a feasible point $\hat{x} \in Q$, we want to bound

$$\delta(\hat{x}) := f(\hat{x}) - f^*. \tag{2.3}$$

As $\hat{x}$ is feasible, this quantity is evidently non-negative. If we allow $\hat{x}$ to take values outside of Q, the difference $\delta(\hat{x})$ might become negative.

We compute an approximate solution to Problem (2.1) by running a piece of code on the idealized real arithmetic computer. Formally, we call this code a **solution method** M for the family $\mathcal{P}$ of Optimization Problems (2.1). In order to solve an instance $p \in \mathcal{P}$, the computer reads in Data(p) and ϵ, performs a finite number of elementary operations instructed by the solution method M, and outputs an ϵ-solution for p.

Let us now discuss the efficiency of the solution method M. We write $\mathrm{Compl}_{\mathrm{M}}(p, \epsilon)$ for the number of elementary operations that are carried out by the computer when we run M for finding an ϵ-solution to problem instance p. This notion involves both the particular problem instance $p \in \mathcal{P}$ and the chosen error tolerance ϵ. However, efficiency of a solution method should be defined not only with respect to a single problem instance, but rather with respect to the whole family of problems. Let us formalize this requirement.

Definition 2.3 (Polynomial-time solution method) *We call* M *a polynomial-time solution method on $\mathcal{P}$ if there exists a polynomial $q : \mathbb{R} \times \mathbb{R} \to \mathbb{R}$ such that:*

$$\mathrm{Compl}_{\mathrm{M}}(p, \epsilon) \leq q\left(\mathrm{Size}(p), \ln\left(\frac{\mathrm{Size}(p) + \|\mathrm{Data}(p)\|_1 + \epsilon^2}{\epsilon}\right)\right) \qquad (2.4)$$

for any $p \in \mathcal{P}$ and for any $\epsilon > 0$.

As discussed in Section 5.1.2 in [BTN01], we can interpret

$$\mathrm{Digits}(p, \epsilon) := \ln\left(\frac{\mathrm{Size}(p) + \|\mathrm{Data}(p)\|_1 + \epsilon^2}{\epsilon}\right)$$

as a bound on the number of accuracy digits of an ϵ-solution. When we fix p and assume that ϵ is bounded from above by a constant $\mathcal{O}(1)$, the numerator in the above fraction can be bounded by another constant $\mathcal{O}(1)$, which allows us to write:

$$\mathrm{Digits}(p, \epsilon) = \mathcal{O}(1)\ln\left(\frac{1}{\epsilon}\right) = \mathcal{O}(1)\alpha \text{ with } \epsilon = 10^{-\alpha}.$$

Let us illustrate the relationship between the accuracy and the complexity of a solution method.

Example 2.4 (Adapted from Section 5.1.2 in [BTN01]) *In Sections 4.2 and 6.2, we study Interior-Point (IP) methods [Kar84], which are of polynomial time and have an Upper Bound (2.4) that is linear with respect to $\mathrm{Digits}(p, \epsilon)$. Let us focus now for a moment on the effects of the solution accuracy $\epsilon > 0$ on the complexity result. We assume that $p \in \mathcal{P}$. Then, the complexity estimate for Interior-Point methods can be written in the form:*

$$\mathrm{Compl}_{\mathrm{IP}}(p, \epsilon) = \mathcal{O}(1)\ln\left(\frac{1}{\epsilon}\right) = \mathcal{O}(1)\alpha \text{ with } \epsilon = 10^{-\alpha}. \qquad (2.5)$$

Suppose that there exists another solution method, denoted by **AFO** *and with the complexity estimate:*

$$\mathrm{Compl}_{\mathrm{AFO}}(p, \epsilon) = \mathcal{O}(1)\frac{1}{\epsilon} = \mathcal{O}(1)10^{\alpha} \text{ with } \epsilon = 10^{-\alpha}. \qquad (2.6)$$

That is, **AFO** *may correspond to one of the advanced First-Order methods that will be discussed in Section 3.4 as well as Chapters 9 and 10.*

It was in 2006 when I attended my first course in Optimization. Let us assume that both solution methods **IP** *and* **AFO** *were able to solve p up to an accuracy of two digits in, say, t seconds using a modern computer in 2006. Moore's law says that the computer speed doubles every 18 months, that is, a modern computer of 2012 is 16 times faster than the computer which was used in 2006. Applying solution method* **IP** *to problem instance p and running it on* *a modern computer in 2012 for the same t seconds, we obtain a solution that is correct up to 32 digits. In contrast, with solution method* AFO, *we obtain a solution with only three accuracy digits.*

Finally, we are ready to define computational tractability of optimization problems.

Definition 2.4 (Polynomially solvable family) *We say that a family $\mathcal{P}$ of optimization problems is polynomially solvable if there exists a polynomial-time solution method for $\mathcal{P}$.*

Polynomial solvability, that is, computational tractability, is a purely theoretical concept. But, what does this concept mean for practical tractability of a family of optimization problems? Is a polynomial-time solution method always fast in practice? Do non-polynomial-time solution methods necessarily show a bad performance in practice? Let us discuss these questions with the aid of an example.

Example 2.5 (Ex. 2.1 cont.; see Sect. 5.1.2, 5.4.2, 6.1 in [BTN01])
A polynomially solvable family $\mathcal{P}$ of optimization problems admits, by definition, a polynomial-time solution method M. *However, the number of elementary operations performed by the computer when running solution method* M *for finding an ϵ-solution to a problem instance $p \in \mathcal{P}$ could grow of the order $[\mathrm{Size}(p)]^{3.5}$, or of the order $[\mathrm{Size}(p)]^{10}$, or even worse. So, polynomial-time solution methods are not necessarily fast in practice. The prime example here is the so-called Ellipsoid method [Sho77] (see [BTN01] for a description of this method). The Ellipsoid method is a polynomial-time solution method for Linear Programming. Nevertheless, it is never used to solve linear optimization problems in practice, as it is not able to solve low/medium-scale problems (say about 100 variables) in reasonable time. The most commonly used, and actually in practice highly efficient, method to solve linear optimization methods is the Simplex method, which was introduced in the late forties by Dantzig [Dan63]. However, Klee and Minty [KM72] showed that the Simplex method is a non-polynomial-time solution method. Given these two facts, the usefulness of theoretical complexity considerations was at least questionable for practical computations since the late seventies / early eighties: it had become an important issue to clarify the existence of polynomial-time solution methods that are also efficient in practice. Finally, Interior-Point methods [Kar84] were introduced for Linear Programming. These methods run indeed in polynomial time and are at the same time highly efficient in practice. The development of Interior-Point methods was a milestone in the history of real arithmetic complexity theory, as it finally showed that theoretical complexity considerations are linked to practical tractability. Nevertheless, we should keep in mind that a solution*

method should not only be analyzed theoretically, but also thoroughly tested in practice. In accordance to this observation, we will perform both a theoretical complexity study and solid numerical tests for all methods presented in this thesis.

Unfortunately, there exist many problems in the general form of (2.1) for which no efficient solution methods are known; see the preface in [BTN01] and Example 1.1.4 in [Nes04] for some examples. This is bad news. The good news is: as we show in Section 2.4, convex optimization problems, that is, a subfamily of optimization problems, which includes among others all linear programs, are computationally tractable under very mild assumptions. First, let us now introduce this subfamily of optimization problems very carefully. Together with convexity, we study some other important properties of the feasible set and the objective function.

2.3 What is a convex optimization problem?

2.3.1 Convexity and other properties of the feasible set

Throughout the thesis, we assume that Q is a subset of $\mathbb{R}^n$ with $1 \leq n < \infty$. We consider $\mathbb{R}^n$ together with the standard Euclidean scalar product, which we denote by $\langle \cdot, \cdot \rangle$. The space $\mathbb{R}^n$ is equipped with a norm $\|\cdot\|$, which may differ from the norm that is induced by the scalar product.

A set is convex if the whole line segment of any two points in this set lies also in the set. More formally:

Definition 2.5 (Convex set) *We say that Q is convex if $\lambda x + (1-\lambda)y \in Q$ for any $x, y \in Q$ and for any $\lambda \in [0, 1]$.*

The following set is of particular interest in this thesis:

Definition 2.6 (Simplex) *We denote by Δ_n the $(n-1)$-dimensional simplex:*

$$\Delta_n := \left\{ x \in \mathbb{R}^n : x \geq 0, \ \sum_{i=1}^{n} x_i = 1 \right\}.$$

Lemma 2.1 Δ_n *is convex.*

Let us briefly recall some other basic definitions.

Definition 2.7 (Interior of a set) *A point $x \in Q$ is called an interior point of Q if there exists an $\epsilon > 0$ such that $\{y \in \mathbb{R}^n : \|y - x\| \leq \epsilon\} \subset Q$. The set $\mathrm{int}(Q)$ of all interior points of Q is called the interior of Q.*

A standard result in Analysis says that all norms defined on $\mathbb{R}^n$ are equivalent. It implies that the above definition is independent of the particular choice of the norm $\|\cdot\|$.
We observe that the above definition concerns the interior of Q with respect to the full space $\mathbb{R}^n$. In some situations, it is more meaningful to study the interior of Q with respect to its **affine hull**, that is, with respect to the set

$$\mathrm{aff}(Q) := \left\{ \sum_{i=1}^n \theta_i x_i : \sum_{i=1}^n \theta_i = 1,\ x_i \in Q,\ \theta_i \in \mathbb{R}\ \forall\ i = 1, \ldots, n \right\}.$$

Definition 2.8 (Relative interior of a set) *We say that* $x \in Q$ *belongs to the relative interior of* Q *if there exists an* $\epsilon > 0$ *such that* $\{y \in \mathbb{R}^n : \|y - x\| \le \epsilon\} \cap \mathrm{aff}(Q) \subset Q$. *We write* $\mathrm{relint}(Q)$ *for the relative interior of* Q.

As for the interior of Q, the definition of the relative interior of Q is independent of the particular choice of the norm $\|\cdot\|$. The following example shows that the sets $\mathrm{int}(Q)$ and $\mathrm{relint}(Q)$ may significantly differ.

Example 2.6 (Simplex) *The set* Δ_n *has an empty interior, but:*

$$\mathrm{relint}(\Delta_n) = \left\{ x \in \mathbb{R}^n : x > 0,\ \sum_{i=1}^n x_i = 1 \right\} \neq \emptyset.$$

Let us continue with the review of some basic definitions.

Definition 2.9 (Open set) *We say that* Q *is an open subset in* $\mathbb{R}^n$ *if the equation* $Q = \mathrm{int}(Q)$ *holds.*

Definition 2.10 (Closed set) *If the complement* $\mathbb{R}^n \setminus Q$ *is an open subset in* $\mathbb{R}^n$, *we call* Q *a closed subset of* $\mathbb{R}^n$.

Definition 2.11 (Compact set) *A set* $Q \subset \mathbb{R}^n$ *is compact if every sequence in* Q *has a subsequence that converges to an element in* Q.

Compact sets are characterized by closed- and boundedness:

Theorem 2.1 (Heine-Borel's Theorem; Theorem 3.5 in [AE05]) *A set* $Q \subset \mathbb{R}^n$ *is compact if and only if it is closed and bounded.*

Example 2.7 (Simplex) *The set* Δ_n *is compact, as it is evidently bounded and closed.*

2.3.2 Objective function: convexity and other properties

Convexity of Optimization Problem (2.1) concerns both the feasible set and the objective function. In this section, we turn our attention now to the objective function.

Throughout this section, we mostly consider functions that are of the form $f : \mathbb{R}^n \to \mathbb{R} \cup \{+\infty\}$. As a first definition, we introduce the domain of f, that is, the set of all $x \in \mathbb{R}^n$ for which $f(x)$ is finite.

Definition 2.12 (Domain) *We refer to the set*

$$\operatorname{dom} f := \{x \in \mathbb{R}^n : f(x) < +\infty\}$$

as the domain of $f : \mathbb{R}^n \to \mathbb{R} \cup \{+\infty\}$.

We always suppose that $\operatorname{dom} f \neq \emptyset$. If f is given by the Optimization Problem (2.1), this requirement is satisfied, as we have assumed that there exists an optimal solution x^* with $f(x^*) < +\infty$.

2.3.2.1 A geometric definition of convex functions

We can use the convexity definition of sets from above in order to specify convex functions. For this, we identify the function $f : \mathbb{R}^n \to \mathbb{R} \cup \{+\infty\}$ as a set in $\mathbb{R}^{n+1}$, which we call the epigraph of f.

Definition 2.13 (Epigraph) *The epigraph of $f : \mathbb{R}^n \to \mathbb{R} \cup \{+\infty\}$ is defined as the set*

$$\operatorname{epi} f := \{(x,t) \in \operatorname{dom} f \times \mathbb{R} : f(x) \leq t\} \subset \mathbb{R}^{n+1}.$$

A graphical illustration of the set $\operatorname{epi} f$ is given in Figure 2.1. This set representation of f can be used to define convex functions.

Definition 2.14 (Convex function) *We say that $f : \mathbb{R}^n \to \mathbb{R} \cup \{+\infty\}$ is convex if $\operatorname{epi} f$ is a convex set.*

Given a function $f : Q \to \mathbb{R}$, we say that h is convex if its extension

$$\bar{f} : \mathbb{R}^n \to \mathbb{R} \cup \{+\infty\} : x \mapsto \begin{cases} f(x), & \text{if } x \in Q \\ +\infty, & \text{otherwise} \end{cases}$$

is convex. On the other hand, a function $h : \mathbb{R}^n \to \mathbb{R} \cup \{+\infty\}$ is called convex on Q if $\operatorname{dom} h \cap Q \neq \emptyset$ and if

$$\overline{h|_Q} : \mathbb{R}^n \to \mathbb{R} \cup \{+\infty\} : x \mapsto \begin{cases} h(x), & \text{if } x \in Q \\ +\infty, & \text{otherwise} \end{cases}$$

is convex. We are positioned to define convex optimization problems.

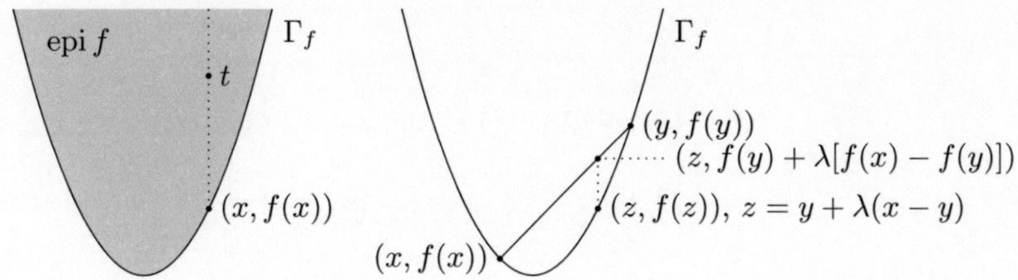

Figure 2.1: *Different characterizations of convex functions. Left figure: geometric characterization by the epigraph. Right figure: analytic characterization by Inequality (2.7). In both figures, we denote by $\Gamma_f := \{(x, f(x)) : x \in \mathbb{R}^n\}$ the graph of the function $f : \mathbb{R}^n \to \mathbb{R} \cup \{+\infty\}$.*

Definition 2.15 (Convex optimization problem) *We say that Problem (2.1) is convex if both the feasible set Q and the objective function f are convex.*

2.3.2.2 An analytic approach to convex functions

In literature, there exist many alternative ways to define convex functions, and from each definition we gain new insights on them. In the next two paragraphs, we display some alternative convexity representations that are relevant for this thesis. For a more complete list and more details, we refer to the books of Rockafellar [Roc70], Nesterov [Nes04], as well as Boyd and Vandenberghe [BV04].

We first give an analytic description of convex functions.

Theorem 2.2 (Theorem 3.1.2 in [Nes04]) *Let $f : \mathbb{R}^n \to \mathbb{R} \cup \{+\infty\}$. The function f is convex if and only if $\operatorname{dom} f$ is convex and if*

$$f(\lambda x + (1 - \lambda)y) \leq \lambda f(x) + (1 - \lambda)f(y) \tag{2.7}$$

for all $x, y \in \operatorname{dom} f$ and for any $\lambda \in [0, 1]$.

For a graphical illustration of the above condition, we refer to Figure 2.1.

2.3.2.3 A dual view on convex functions

Finally, let us describe convex functions by affine mappings.

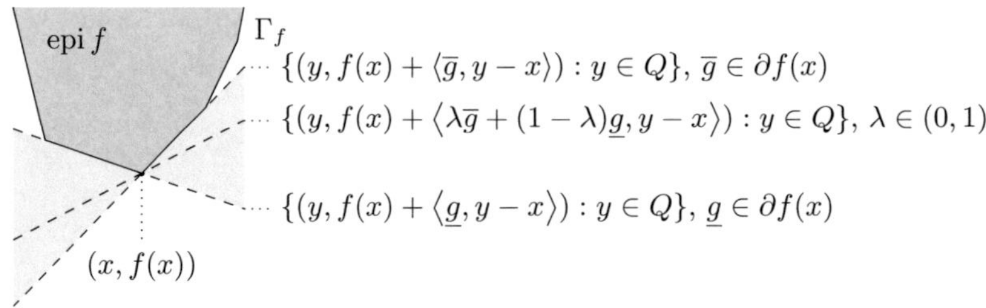

Figure 2.2: *The dark gray zone corresponds to the epigraph of a convex function $f : \mathbb{R}^n \to \mathbb{R} \cup \{+\infty\}$. In light gray, we show the set of all affine spaces passing through $(x, f(x))$, $x \in \mathbb{R}^n$, and defining half-spaces that contain the entire epigraph of f. By Γ_f we denote the graph of f.*

Definition 2.16 (Subgradient) *Let $f : \mathbb{R}^n \to \mathbb{R} \cup \{+\infty\}$ be a convex function. We say that $g \in \mathbb{R}^n$ is a subgradient of f at $x \in \operatorname{dom} f$ if*

$$f(y) \geq f(x) + \langle g, y - x\rangle \qquad \forall\, y \in \operatorname{dom} f.$$

The set of all subgradients of f at $x \in \operatorname{dom} f$ is called the subdifferential of f at x and denoted by $\partial f(x)$.

Remark 2.1 *Every element $g \in \partial f(x)$ gives rise to an affine mapping*

$$y \mapsto s(y) := f(x) + \langle g, y - x\rangle$$

from $\mathbb{R}^n$ to $\mathbb{R}$, where its linear part represents a dual element through the scalar product. According to the Riesz Representation Theorem, this representation is unique (see for instance Theorem 2.14 in [AE08]).

The subdifferential $\partial f(x)$ of the convex function f at $x \in \operatorname{dom} f$ can be associated with affine spaces that have the following two characteristics:

1. these affine spaces lie in $\mathbb{R}^{n+1}$ and pass through $(x, f(x))$;

2. they contain the entire epigraph of f;

see Figure 2.2 for a graphical illustration.
Let us discuss the existence of subgradients.

Theorem 2.3 (Theorem 23.4 in [Roc70]) *Let $f : \mathbb{R}^n \to \mathbb{R} \cup \{+\infty\}$ be a convex function. We have:*

1. *For any $x \in \text{relint}(\text{dom } f)$, the set $\partial f(x)$ is non-empty.*

2. *If $x \notin \text{dom } f$, the set $\partial f(x)$ is empty.*

Vice versa, we can show:

Theorem 2.4 (Lemma 3.1.6 in [Nes04]) *Let $f : \mathbb{R}^n \to \mathbb{R} \cup \{+\infty\}$. The function f is convex if $\text{dom } f$ is convex and if*

$$\{g \in \mathbb{R}^n : f(y) \geq f(x) + \langle g, y - x \rangle \ \forall \, y \in \text{dom } f\} \neq \emptyset \qquad \forall \, x \in \text{dom } f.$$

We conclude this subsection by presenting some results that indicate the importance of subgradients in Convex Optimization.

Theorem 2.5 (Theorem 3.1.15 in [Nes04]) *Let $f : \mathbb{R}^n \to \mathbb{R} \cup \{+\infty\}$ be a convex function and $x^* \in \text{dom } f$. We have $f(x^*) \leq f(y)$ for any $y \in \text{dom } f$ if and only if $(0, \ldots, 0)^T \in \partial f(x^*)$.*

Theorem 2.6 (Theorem 25.1 in [Roc70]) *Let $f : \mathbb{R}^n \to \mathbb{R} \cup \{+\infty\}$ be a convex function and $x \in \text{dom } f$. If the function f is differentiable at x, then $\partial f(x) = \{\nabla f(x)\}$. Vice versa, the function f is differentiable at x if it has a unique subgradient at x.*

Theorem 2.7 (Theorem 27.4 in [Roc70]) *Let $f : \mathbb{R}^n \to \mathbb{R} \cup \{+\infty\}$ be a convex function and $x^* \in \text{dom } f$. It holds that $f(x^*) \leq f(y)$ for any $y \in \mathbb{R}^n$ if and only if there exists an element $g \in \partial f(x^*)$ for which $\langle g, x - x^* \rangle \geq 0$ for any $x \in \text{relint}(\text{dom } f)$.*

2.3.2.4 Strongly convex functions

According to Theorem 2.3, any convex function can be approximated from below by affine functions (on the relative interior of its domain). However, these underestimators do not include any information about the curvature of the convex function. Note that Theorem 2.3 cannot be improved, as the convex function might not have any curvature at all; see Figure 2.2 for an example. However, many convex functions allow us to bound their curvature from below. These functions are called strongly convex.

Definition 2.17 (Strongly convex function) *We say that the function $f : \mathbb{R}^n \to \mathbb{R} \cup \{+\infty\}$ is strongly convex with modulus $\sigma = \sigma(\mathbb{R}^n) > 0$ if $\text{dom } f$ is convex and if*

$$\lambda f(x) + (1 - \lambda) f(y) \geq f(\lambda x + (1 - \lambda) y) + \frac{\sigma}{2} \lambda (1 - \lambda) \|y - x\|^2$$

for all $x, y \in \text{dom } f$ and for any $\lambda \in [0, 1]$.

Note that the parameter $\sigma = \sigma(\mathbb{R}^n)$ depends on the choice of the norm $\|\cdot\|$. We call a function $f : Q \to \mathbb{R}$ strongly convex with modulus $\sigma(Q)$ if its extension

$$\bar{f} : \mathbb{R}^n \to \mathbb{R} \cup \{+\infty\} : x \mapsto \begin{cases} f(x), & \text{if } x \in Q \\ +\infty, & \text{otherwise} \end{cases}$$

is strongly convex with modulus $\sigma(Q)$. In addition, we say that the function $h : \mathbb{R}^n \to \mathbb{R} \cup \{+\infty\}$ is strongly convex with modulus $\sigma(Q)$ on Q if the set $\operatorname{dom} h \cap Q$ is non-empty and if

$$\overline{h|_Q} : \mathbb{R}^n \to \mathbb{R} \cup \{+\infty\} : x \mapsto \begin{cases} h(x), & \text{if } x \in Q \\ +\infty, & \text{otherwise} \end{cases}$$

is strongly convex modulus $\sigma(Q)$. We often make a slight notational abuse and write σ instead of $\sigma(Q)$.

As for convex functions, we can give a characterization of strongly convex functions using first-order information.

Theorem 2.8 (Theorem VI.6.1.2 in [HUL93]) [3]   *Assume   that $f : \mathbb{R}^n \to \mathbb{R} \cup \{+\infty\}$ is a convex function. The function f is strongly convex with modulus $\sigma > 0$ if and only if*

$$f(y) \geq f(x) + \langle g, y - x \rangle + \frac{\sigma}{2} \|y - x\|^2 \qquad \forall\, x, y \in \operatorname{dom} f, \ \forall\, g \in \partial f(x).$$

As the curvature is bounded from below, strongly convex functions have a unique minimizer over the set Q, provided that this set is closed and convex.

Theorem 2.9 (Lemma 6 in [Nes09]) *Consider Problem (2.1) with a closed and convex feasible set $Q \subset \operatorname{dom} f$. If f is strongly convex and continuous on Q, Optimization Problem (2.1) is solvable, its optimal solution x^* is unique, and it holds that*

$$f(x) \geq f(x^*) + \frac{\sigma}{2} \|x - x^*\|^2 \qquad \forall\, x \in Q.$$

We conclude this subsection by presenting several examples of strongly convex functions.

Example 2.8 (Euclidean setup; see [Nes09]) *Consider the space $\mathbb{R}^n$ together with the Euclidean norm. The function*

$$d_{euc}(x) : \mathbb{R}^n \to \mathbb{R}_{\geq 0} : x \mapsto \frac{1}{2} \|x\|_2^2$$

is strongly convex with modulus 1.

---

[3]Theorem VI.6.1.2 in [HUL93] explores the necessary and sufficient conditions under which a real-valued convex function $f : \mathbb{R}^n \to \mathbb{R}$ is strongly convex on a convex set Q. The same arguments as utilized in [HUL93] also show the correctness of Theorem 2.8.

The following example is of particular interest in the context of this thesis.

Example 2.9 (Simplex setup; see [Nes09]) *We equip $\mathbb{R}^n$ with the norm $\|x\|_1 := \sum_{i=1}^n |x_i|$. The shifted negative entropy function*

$$d_{\Delta_n}(x) : \Delta_n \to \mathbb{R}_{\geq 0} : x \mapsto \ln(n) + \sum_{i=1}^n x_i \ln(x_i)$$

is strongly convex with modulus 1.

Let us quickly comment on the choice of the 1-norm. This comment involves the dual norm.

Definition 2.18 (Dual norm) *The dual norm $\|\cdot\|_*$ to $\|\cdot\|$ is defined as*

$$\|u\|_* := \max_{x \in \mathbb{R}^n} \left\{ \langle u, x \rangle : \|x\| = 1 \right\}, \qquad u \in \mathbb{R}^n.$$

Clearly, we have $\|x\|_1 \geq \|x\|_2$ for any $x \in \mathbb{R}^n$. However, the dual norm $\|\cdot\|_\infty$ to $\|\cdot\|_1$ satisfies the converse inequality. We recall that the dual of the Euclidean norm is the Euclidean norm itself.

The above simplex setting can be naturally extended to matrices.

Example 2.10 (Matrix simplex setup; see [Nes07]) *We consider the space $\mathcal{S}^n$ of symmetric real $(n \times n)$-matrices. Given a matrix $X \in \mathcal{S}^n$, we write $\lambda_n(X) \geq \ldots \geq \lambda_1(X)$ for its eigenvalues and equip $\mathcal{S}^n$ with the norm $\|X\|_{(1)} := \sum_{i=1}^n |\lambda_i(X)|$. The set*

$$\Delta_n^M := \left\{ X \in \mathcal{S}^n : \lambda_1(X) \geq 0, \ \sum_{i=1}^n \lambda_i(X) = 1 \right\}$$

*is called the **simplex in matrix form**. When we restrict this set to diagonal matrices, we recover Δ_n. By applying the shifted negative entropy function d_{Δ_n} to the eigenvalues of $X \in \Delta_n^M$, we can define the function*

$$d_{\Delta_n^M}(X) : \Delta_n^M \to \mathbb{R}_{\geq 0} : X \mapsto \ln(n) + \sum_{i=1}^n \lambda_i(X) \ln(\lambda_i(X)).$$

As shown in [Nes07], this function is strongly convex with modulus 1.

2.3.2.5 Differentiable functions with Lipschitz continuous gradient

According to the last two subsections, we can approximate any convex function from below by an affine function, or, in the case of strongly convex functions, by a function that is quadratic with respect to the norm $\|\cdot\|$. In other words, we approximate the epigraph of a function from outside, that is, with a larger set. In this subsection, we consider the opposite approximation problem: we derive an approximation of the epigraph from inside.

It turns out that we can give such an approximation for differentiable functions with a Lipschitz continuous gradient. The definition of this class of functions includes the norm of the gradients. The gradient represents a dual element through the scalar product. Consequently, we need to use the dual norm for the gradients.

Definition 2.19 (Lipschitz continuous gradient) *Assume that $Q \subset U$ for an open set $U \subset \mathbb{R}^n$. Let $f : \mathbb{R}^n \to \mathbb{R} \cup \{+\infty\}$ be a function which is differentiable on U. We say that f has a Lipschitz continuous gradient on Q if there exists a constant $L(Q) > 0$ such that:*

$$\|\nabla f(x) - \nabla f(y)\|_* \leq L(Q) \, \|x - y\| \qquad \forall \, x, y \in Q. \tag{2.8}$$

We refer to $L(Q)$ as the Lipschitz constant of the gradient of f on Q. We write $f \in C^1_{L(Q)}$ if f is differentiable on an open set that includes Q and if the gradient of f satisfies (2.8).

We can give the following characterization of differentiable functions with Lipschitz continuous gradients. If $f \in C^1_{L(Q)}$, we can use this characterization to derive an approximation of epi f from inside.

Theorem 2.10 (Theorem 2.1.5 in [Nes04]) *Let $U \subset \mathbb{R}^n$ be an open set with $Q \subset U$ and $f : \mathbb{R}^n \to \mathbb{R} \cup \{+\infty\}$ be a function that is differentiable on U. This function belongs to $C^1_{L(Q)}$ if and only if*

$$f(y) \leq f(x) + \langle \nabla f(x), y - x \rangle + \frac{L(Q)}{2} \, \|x - y\|^2 \qquad \forall \, x, y \in Q.$$

If there is no possibility of confusion, we abbreviate $L(Q)$ into L.

2.4 Convex optimization problems are polynomially solvable!

"In fact the great watershed in Optimization is not between linearity and non-linearity, but convexity and non-convexity."

(R. Tyrell Rockafellar, 1993; see Page 185 in [Roc93])

Let us now review the assumptions under which convex optimization problems become computationally tractable. We follow closely Section 5.3 in the book of Ben-Tal and Nemirovski [BTN01]. We assume that $\mathcal{P}$ is a family of optimization problems in the form of

$$\min_{x \in Q} f(x), \tag{2.9}$$

where both $Q \subset \mathbb{R}^n$ and $f : \mathbb{R}^n \to \mathbb{R}$ are convex. We equip the family $\mathcal{P}$ with an infeasibility measure $\mathrm{Infeas}_{\mathcal{P}}(\cdot, \cdot)$.

Definition 2.20 (Polynomially computable family) *We call $\mathcal{P}$ polynomially computable if there are two codes C_1 and C_2 for the real arithmetic computer. These codes satisfy for any $p \in \mathcal{P}$, any $x \in \mathbb{R}^n$, and any $\epsilon > 0$:*

1. for input $\mathrm{Data}(p)$ and x, the code C_1 returns the objective function value $f(x)$ and a subgradient of f at x. The running time of C_1 is bounded from above by a polynomial in $\mathrm{Size}(p)$;

2. for input $\mathrm{Data}(p)$, x, and $\epsilon > 0$, the code C_2 outputs whether $\mathrm{Infeas}_{\mathcal{P}}(p, x) \leq \epsilon$ or not. In the later case, it returns a hyperplane which separates x from all $y \in \mathbb{R}^n$ that satisfy $\mathrm{Infeas}_{\mathcal{P}}(p, y) \leq \epsilon$. The running time of this code is bounded from above by a polynomial in $\mathrm{Size}(p)$ and $\mathrm{Digits}(p, \epsilon)$.

Example 2.11 (Ex. 2.1 continued; see Section 5.3 in [BTN01])
The family $\mathcal{LP}$ of linear optimization problems is polynomially solvable, as:

1. for any $x \in \mathbb{R}^n$, the computation of the objective function value $c^T x$ requires $\mathcal{O}(n)$ arithmetic operations. The gradient of the objective function at x is c;

2. for any $x \in \mathbb{R}^n$ and any $p \in \mathcal{LP}$, the computation of

$$\mathrm{Infeas}_{\mathcal{LP}}(x, p) = \max \left[0, \max_{1 \leq j \leq m} \{(Ax)_j - b_j\}, \max_{1 \leq i \leq n} \{-x_i\} \right]$$

needs $\mathcal{O}(mn)$ arithmetic operations. If $\mathrm{Infeas}_{\mathcal{LP}}(x, p) > \epsilon$ for a fixed $\epsilon > 0$, then there exists $1 \leq i^* \leq n$ or $1 \leq j^* \leq m$ such that $x_{i^*} < -\epsilon$ or such that $(Ax)_{j^*} - b_{j^*} > \epsilon$. Let us focus on the second case, as the first case works similarly. The j^*-th row of A defines a hyperplane with $(Ax)_{j^*} > b_{j^*} + \epsilon$ and such that

$$(Ay)_{j^*} \leq b_{j^*} + \epsilon \qquad \forall\, y \in \mathbb{R}^n \text{ with } \mathrm{Infeas}_{\mathcal{LP}}(y, p) \leq \epsilon.$$

Definition 2.21 (Family with polynomial growth) *We say that $\mathcal{P}$ is of polynomial growth if there exists a polynomial $q : \mathbb{R} \to \mathbb{R}$ such that:*

$$|f(x)| + \mathrm{Infeas}_{\mathcal{P}}(p, x) \leq \mathcal{O}(1) \left(\mathrm{Size}(p) + \|x\|_1 + \|\mathrm{Data}(p)\|_1\right)^{q(\mathrm{Size}(p))}$$

for any $x \in \mathbb{R}^n$ and any $p \in \mathcal{P}$.

Clearly, the family $\mathcal{LP}$ of linear optimization problems is of polynomial growth.

Definition 2.22 (Polynomially bounded feasible sets) *The family $\mathcal{P}$ of optimization problems is with polynomially bounded feasible sets if there exists a polynomial $q : \mathbb{R} \to \mathbb{R}$ such that:*

$$Q = Q_p \subset \left\{ x \in \mathbb{R}^n : \|x\|_2 \leq \mathcal{O}(1) \left(\mathrm{Size}(p) + \|\mathrm{Data}(p)\|_1\right)^{q(\mathrm{Size}(p))} \right\}$$

for any $p \in \mathcal{P}$, where $Q = Q_p$ denotes the feasible set of p.

We have prepared all the necessary conditions for computational tractability.

Theorem 2.11 (Theorem 5.3.1 in [BTN01]) *The family $\mathcal{P}$ of Optimization Problems (2.9) is polynomially solvable if $\mathcal{P}$ is polynomially computable, of polynomial growth, and with polynomially bounded sets.*

Linear and semidefinite optimization problems are polynomially computable and with polynomial growth; see the previous examples and Section 5.3 in [BTN01] for the details. However, these families of optimization problems do not necessarily have polynomially bounded feasible sets. Fortunately, as shown in Section 5.3 of [BTN01], there is a brute-force trick to invoke this property with no restrictions for all practical computations. We conclude:

Corollary 2.2 (Section 5.3 in [BTN01]) *The families of all linear and semidefinite optimization problems that appear in practice are polynomially solvable.*

2.5 Bregman distances

Let us now focus again on the feasible set. We assume that Q is a closed and convex set. In addition, we choose a reference point $x_0 \in \mathrm{relint}(Q)$. A priori, we measure the distance of any $y \in Q$ with respect to the chosen reference point using the Euclidean metric. However, there is no reason to believe that this metric is well adapted to the geometry of the feasible set. In this section, we introduce Bregman distances. Provided that we initialize Bregman distances appropriately, we can use it to reproduce the geometry of the sets Δ_n and Δ_n^M – the most relevant feasible sets of this thesis – in a much more accurate way than with the standard Euclidean metric.

First, we need to introduce the notion of distance-generating functions.

Definition 2.23 (Distance-generating function) *Let $Q \subset \mathbb{R}^n$ be closed and convex. We say that the function $d\colon Q \to \mathbb{R}$ is a distance-generating function if the following three conditions hold:*

1. *d is continuous on Q;*

2. *d is strongly convex with modulus 1 on Q;*

3. *given the set $Q^o(d) := \{x \in Q : \partial d(x) \neq \emptyset\}$, there exists a continuous mapping $d' : Q^o(d) \to \mathbb{R}^n : x \mapsto d'(x) := g(x)$, where $g(x) \in \partial d(x)$.*

If there is no possibility for confusion, we write Q^o instead of $Q^o(d)$.

Example 2.12 (Ex. 2.8 - 2.10 continued; see [Nes07, Nes09])
The functions d_{euc}, d_{Δ_n}, and $d_{\Delta_n^M}$ are distance-generating functions. For these functions, we obtain the sets $Q^o(d_{euc}) = \mathbb{R}^n$, $Q^o(d_{\Delta_n}) = \mathrm{relint}(\Delta_n)$, and $Q^o(d_{\Delta_n^M}) = \mathrm{relint}(\Delta_n^M)$. Note that these functions are continuously differentiable on their corresponding sets Q^o.

Let $d : Q \to \mathbb{R}$ be a distance-generating function defined on the closed and convex set Q. In this thesis, we study several algorithms for solving convex optimization problems. The following type of convex optimization problems needs to be routinely solved in these methods:

$$\min_{x \in Q} \left\{ \phi_s(x) := d(x) - \langle s, x \rangle \right\}, \qquad s \in \mathbb{R}^n. \tag{2.10}$$

Let $s \in \mathbb{R}^n$. Due to the strong convexity of the distance-generating function d, the function $\phi_s(x) := d(x) - \langle s, x \rangle$ is also strongly convex on Q. Applying Theorem 2.9, we observe that Problem (2.10) has a unique minimizer x_s^*. By its definition, this minimizer satisfies $\phi_s(x) \geq \phi_s(x_s^*)$ for any $x \in Q$, which can be rewritten as:

$$d(x) \geq d(x_s^*) + \langle s, x - x_s^* \rangle \qquad \forall\, x \in Q.$$

In particular, the element s belongs to $\partial d(x_s^*)$, and thus $x_s^* \in Q^o$. We have verified the following result.

Remark 2.2 *Problem (2.10) has a unique minimizer and this minimizer is in Q^o.*

The practical success of the methods which we study in this thesis depends heavily on the effort that is required to find this unique minimizer. We make therefore the following assumption, which shall hold for the first part of this thesis excluding Section 4.2.

Assumption 2.1 *The unique minimizer x_s^* of (2.10) can be written in a closed form.*

Let us verify that the distance-generating functions d_{euc}, d_{Δ_n}, and $d_{\Delta_n^M}$ satisfy this requirement.

Example 2.13 (Euclidean setup; see [Nes09]) *Given $s \in \mathbb{R}^n$, we obtain:*

$$s = \arg \min_{x \in \mathbb{R}^n} \{d_{euc}(x) - \langle s, x \rangle\}.$$

Example 2.14 (Simplex setup; see [Nes09]) *For $s \in \mathbb{R}^n$, we have:*

$$\left[\frac{\exp(s_i)}{\sum_{k=1}^n \exp(s_k)} \right]_{i=1}^n = \arg \min_{x \in \Delta_n} \{d_{\Delta_n}(x) - \langle s, x \rangle\},$$

whose computation requires $\mathcal{O}(n)$ arithmetic operations.

Example 2.15 (Matrix simplex setup; see [Nem04a, Nes07]) *Given two real $(n \times n)$-matrices A and B, we write*

$$\langle A, B \rangle_F := \sum_{i,j}^n A_{ij} B_{ij}$$

*for the **Frobenius scalar product**. Let $X \in \mathcal{S}^n$ be a symmetric real $(n \times n)$-matrix. We denote by $\lambda(X)$ the vector of eigenvalues of the matrix X. Given a vector $\lambda \in \mathbb{R}^n$, $\mathrm{Diag}(\lambda)$ represents the diagonal matrix whose diagonal is λ. Every $X \in \mathcal{S}^n$ admits an eigendecomposition*

$$X = Q(X)\mathrm{Diag}(\lambda(X))Q(X)^T = \sum_{i=1}^n \lambda_i(X)q_i(X)q_i(X)^T,$$

where $Q(X) := (q_1(X), \ldots, q_n(X))$ is a (not necessarily unique) orthogonal matrix of dimension $n \times n$, and $q_1(X), \ldots, q_n(X)$ form an orthogonal basis of unitary eigenvectors. An eigendecomposition of a generic symmetric matrix can be computed in $\mathcal{O}(n^3)$ elementary operations [HJ96].
Let $S \in \mathcal{S}^n$ with eigendecomposition $S = Q(S)\mathrm{Diag}(\lambda(S))Q(S)^T$. Then,

$$Q(S)\mathrm{Diag}(\lambda^*(S))Q(S)^T = \arg \min_{X \in \Delta_n^M} \left\{ d_{\Delta_n^M}(X) - \langle S, X \rangle_F \right\}, \qquad (2.11)$$

where

$$\lambda_i^*(S) := \frac{\exp(\lambda_i(S))}{\sum_{k=1}^n \exp(\lambda_k(S))} \qquad \forall\, i = 1, \ldots, n.$$

The computation of the minimizer to the optimization problem defined in (2.11) requires $\mathcal{O}(n^3)$ arithmetic operations.

Applying Remark (2.2) for $s = 0$, we can define the d-center.

Definition 2.24 (d-center) *The d-center is defined as*

$$c(d) := \arg\min_{x \in Q} d(x) \in Q^o.$$

Without loss of generality, we assume that the distance-generating function d vanishes at its d-center.

Provided that there is no possibility of confusion, we write c instead of $c(d)$. As d vanishes at c, Theorem 2.9 implies the following lower bound on the distance-generating function:

$$d(x) \geq \frac{1}{2} \|x - c\|^2 \qquad \forall\, x \in Q. \tag{2.12}$$

The d-centers of the running examples are very well-known.

Example 2.16 (Ex. 2.8 - 2.10 cont.; see [Nem04a, Nes07, Nes09])
We denote by $\mathbb{I}_n$ the identity matrix in $\mathbb{R}^{n \times n}$. On the sets $\mathbb{R}^n$, Δ_n, and Δ_n^M, we have:

$$c(d_{euc}) = (0, \ldots, 0)^T, \;\; c(d_{\Delta_n}) = \frac{1}{n}(1, \ldots, 1)^T, \;\; and \;\; c(d_{\Delta_n^M}) = \frac{1}{n}\mathbb{I}_n,$$

respectively.

Let $d : Q \to \mathbb{R}$ be a distance-generating function. Given this function, we can define Bregman distances on Q.

Definition 2.25 (Bregman distance) *Let $x \in Q^o$ and $y \in Q$. The Bregman distance of y with respect to x is defined by the function*

$$V_x : Q \to \mathbb{R}_{\geq 0} : y \mapsto V_x(y) := d(y) - d(x) - \langle d'(x), y - x \rangle.$$

See Figure 2.3 for a graphical illustration of this definition. Note that the Bregman distance does not define a metric on Q, as neither it is symmetric, nor it fulfills the triangle inequality.
We deduce the Bregman distances for the distance-generating functions that we introduced in Examples 2.8 - 2.10.

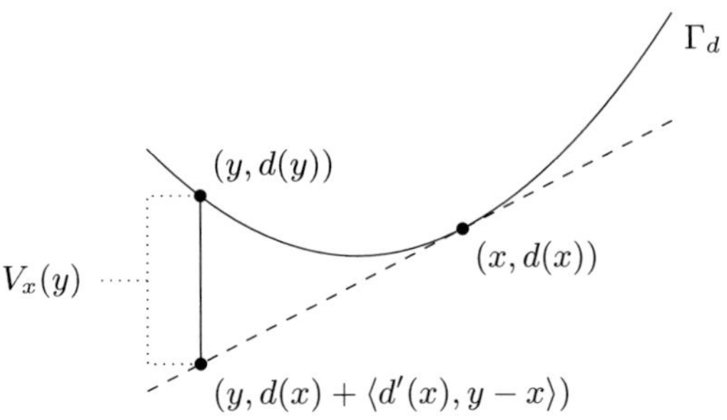

Figure 2.3: *Bregman distance $V_x(y)$. The graph of the distance-generating function d is denoted by Γ_d.*

Example 2.17 (Ex. 2.8 - 2.10 cont.; see [Nem04a, Nes07, Nes09])
For the three functions d_{euc}, d_{Δ_n}, and $d_{\Delta_n^M}$, we obtain the following Bregman distances:

1. $\boldsymbol{d_{euc}}$: *With $x, y \in \mathbb{R}^n$, we have $V_x(y) = \frac{1}{2} \|y - x\|_2^2$.*

2. $\boldsymbol{d_{\Delta_n}}$: *Given $x \in \operatorname{relint}(\Delta_n)$, it holds that:*

$$V_x(y) = \sum_{i=1}^n y_i \ln\left(\frac{y_i}{x_i}\right) \qquad \forall\, y \in \Delta_n.$$

   *In Probability Theory and in Machine Learning, this quantity is referred to as **Kullback-Leibler divergence** and **relative entropy**, respectively.*

3. $\boldsymbol{d_{\Delta_n^M}}$: *We obtain:*

$$V_X(Y) = \sum_{i=1}^n \lambda_i(Y) \ln\left(\frac{\lambda_i(Y)}{\lambda_i(X)}\right) \qquad \forall\, X \in \operatorname{relint}(\Delta_n^M),\ \forall\, Y \in \Delta_n^M.$$

In Figure 2.4, we show the level sets for the Bregman distances defined by d_{euc} and by d_{Δ_3}, respectively, on the two-dimensional simplex Δ_3. We observe that the relative entropy reflects the geometry of the set Δ_3 much more accurately than the squared Euclidean norm.

Let $x \in Q^o$ and consider the Bregman distance $V_x(\cdot)$ defined by d:

$$V_x : Q \to \mathbb{R}_{\geq 0} : y \mapsto V_x(y) := d(y) - d(x) - \langle d'(x), y - x \rangle.$$

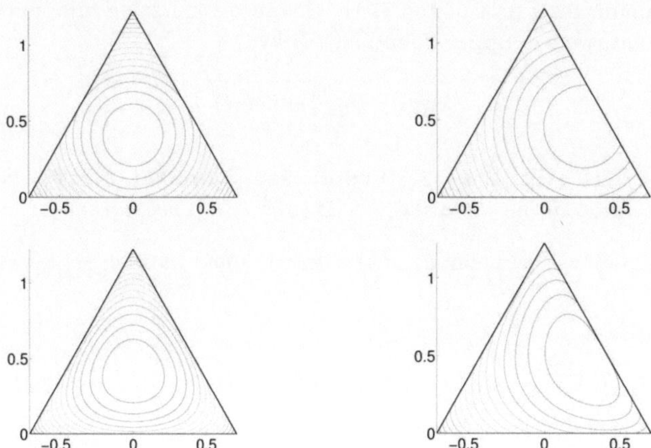

Figure 2.4: *We present the level sets for different Bregman distances $V_x(y) := d(y) - d(x) - \langle d'(x), y - x \rangle$ on the set Δ_3. Top left figure: We set $d = d_{euc}$ and $x = (1/3, 1/3, 1/3)$. Top right figure: $d = d_{euc}$, $x = (1/8, 1/2, 3/8)$. Bottom left figure: $d = d_{\Delta_3}$, $x = (1/3, 1/3, 1/3)$. Bottom right figure: $d = d_{\Delta_3}$, $x = (1/8, 1/2, 3/8)$. We observe that the Bregman distance defined by the entropy function reflects the geometry of the set Δ_3 much more accurately than the squared Euclidean norm.*

Due to the strong convexity of d, we can give an improved lower bound on the Bregman distance:

$$V_x(y) \geq \frac{1}{2} \|x - y\|^2 \qquad \forall\, x \in Q^o,\ \forall\, y \in Q. \tag{2.13}$$

Bregman-distances give rise to the so-called d-diameter of the set Q.

Definition 2.26 (d-diameter) *The d-diameter of the set Q is defined as*

$$\Omega_V(d) := \sqrt{2 \sup_{y \in Q} V_{c(d)}(y)}.$$

Note that the supremum in the above definition can be replaced by a maximum if the feasible set Q is compact. As before, we write Ω_V instead $\Omega_V(d)$ if it is clear from the context which distance-generating function we refer to.

As $c(d)$ minimizes d over Q and as the distance-generating function vanishes at this point, we can upper bound $\Omega_V(d)$ by:

$$\Omega_V(d) \leq \sqrt{2 \sup_{x \in Q} d(x)}.$$

Example 2.18 (Ex. 2.8 - 2.10 cont.; see [Nem04a, Nes07, Nes09]) *In the situation of the three sets $\mathbb{R}^n$, Δ_n, and Δ_n^M, we have:*

$$\sup_{x \in \mathbb{R}^n} d_{euc}(x) = +\infty, \quad \max_{x \in \Delta_n} d_{\Delta_n}(x) = \ln(n), \quad and \quad \max_{X \in \Delta_n^M} d_{\Delta_n^M}(X) = \ln(n),$$

which implies:

$$\Omega_V(d_{euc}) = +\infty, \quad \Omega_V(d_{\Delta_n}) \leq \sqrt{2\ln(n)}, \quad and \quad \Omega_V(d_{\Delta_n^M}) \leq \sqrt{2\ln(n)},$$

respectively.

Let us conclude this subsection by discussing the relation between $\Omega_V(d)$ and the standard diameter of the set Q. Recall that the standard diameter of Q is defined as follows.

Definition 2.27 (Diameter) *We write $\Omega_Q := \sup_{x,y \in Q} \|x - y\|$ for the diameter of Q.*

Applying (2.13), we obtain:

$$\|y - c(d)\| \leq \Omega_V(d) \text{ for any } y \in Q, \text{ and thus } \Omega_Q \leq 2\Omega_V(d). \qquad (2.14)$$

These definitions and observations will be heavily used in the next chapters.

Black-Box optimization methods

Iɴ this and the next chapter, we present several solution methods for solving a broad class of convex optimization problems. All these methods are of **iterative** nature, that is, a set of instructions is repeated until a predefined stopping criterion is satisfied. Together with the methods, we always give a theoretical worst-case analysis of the **analytical complexity**. The analytical complexity corresponds to the number of iterations that need to be performed to meet the stopping criterion. The **arithmetic complexity**, that is, the total number of arithmetic operations that are required by the method until the stopping criterion is fulfilled, is then given by the analytical complexity multiplied by the **cost per iteration** (plus the setup cost of the method); see for instance [Nes04]. While the analytical complexity of a method remains unchanged for the whole family of problems that can be solved by this method, the cost per iteration typically varies a lot depending on the problem's explicit format. The discussion of the iteration cost is thus postponed to Chapters 6 - 10, where we apply these methods to semidefinite optimization problems.

Convex optimization methods can be divided into two classes of methods, namely in **Black-Box optimization methods** and in **Structural optimization methods**; see for instance [NY83, Nes04]. While Structural optimization methods have access to the exact format of the problem they are dealing with, we hide the problem from the method in Black-Box Optimization. In particular, Black-Box optimization methods have access only to a limited amount of information on the problem they are dealing with. In this chapter, we focus on Black-Box optimization methods. Schemes that exploit

Overview of solution methods for finding feasible ϵ-solutions to convex optimization problems

Solution method	Analytical complexity[a]	Optimization framework	Type of information	Discussed in
Primal-Dual Subgradient methods	$\mathcal{O}\left(1/\epsilon^2\right)$	Black-Box	First-Order	Section 3.3
Mirror-Descent methods[b]	$\mathcal{O}\left(1/\epsilon^2\right)$	Black-Box	First-Order	Section 3.3
Smoothing Techniques	$\mathcal{O}\left(1/\epsilon\right)$	Structural	First-Order	Section 4.1.1
Mirror-Prox methods	$\mathcal{O}\left(1/\epsilon\right)$	Structural	First-Order	Section 4.1.2
Optimal First-Order methods	$\mathcal{O}\left(1/\sqrt{\epsilon}\right)$	Black-Box	First-Order	Section 3.4
Interior-Point methods	$\mathcal{O}\left(\ln\left[1/\epsilon\right]\right)$	Structural	Second-Order	Section 4.2

Table 3.1: *Overview of the convex optimization methods that are discussed in this thesis. For each method, we give its analytical complexity for finding a feasible ϵ-solution, we say whether it is a Black-Box or Structural optimization method and whether it is a First- or a Second-Order method, and we refer to the section where we introduce the method.)*

[a]We display here the analytical complexity only with respect to the solution accuracy ϵ. These complexity results typically involve other problem parameters such as the Lipschitz constant of the gradient or an upper bound on the norm of the subgradients. Note that we order the methods according to their analytical complexity; from the slowest to the fastest method.

[b]Mirror-Descent methods are a particular subclass of Primal-Dual Subgradients methods; see Section 3.3. For the sake of completeness, we explicitly mention Mirror-Descent methods here, as well.

the explicit problem structure are left for the next chapter. In Table 3.1, we give an overview of all convex optimization methods that we discuss in this thesis. For each method, we give its analytical complexity with respect to the solution accuracy ϵ and show whether the method belongs to the class of Black-Box or Structural optimization methods. In addition, we indicate for each method whether it is a First-Order or a Second-Order method. A **First-Order method** uses only first-order information on the problem (that is, function values and subgradients of the objective function) to update the iterates, whereas a **Second-Order method** utilizes function values, gradients, and Hessians of the objective function. For reference, we also add the section where we introduce the corresponding method.

We start this chapter by introducing the concept of Black-Box Optimization in Section 3.1. In Sections 3.2 - 3.4, we study several algorithms belonging to this class of optimization algorithms: Dual Averaging schemes (Section 3.2), Mirror-Descent and Primal-Dual Subgradient methods (Section 3.3), as well as Optimal First-Order methods (Section 3.4). The methods are ordered according to their requirements on the objective function. We start with methods that make no assumptions on the objective function (except the existence of an oracle that provides certain information) and add step by step more properties to the objective function, which can be used to strengthen the theoretical guarantees of the methods or to design more elaborated algorithms. Note that Dual Averaging schemes are not listed explicitly in Table 3.1 of convex optimization methods. When we apply Dual Averaging schemes to convex optimization problems, we call the resulting algorithms Primal-Dual Subgradient methods, which can be found in Table 3.1. Interestingly, although being introduced for solving convex optimization problems, Dual Averaging schemes can be applied in a much broader context than just in Convex Optimization. In fact, we exploit this much broader applicability of these schemes in Chapter 5, where we embed an algorithm from Machine Learning in the context of Dual Averaging schemes.

Contributions: In Section 3.4, we present a refined version of the original Optimal First-Order methods introduced in [Nes04]. The original methods use the Lipschitz constant at every instance to update the iterates. We replace this global constant by local estimates, which yields to impressively faster methods in practice; see Chapter 11. The theoretical worst-case bounds on the analytical complexity of the original and the refined methods are of the same order. The methods and complexity results presented in Sections 3.1 - 3.3 are known.

Relevant references: We mainly use the references [NY83, Nes04, Nes05, Nes09]. Parts of this chapter are taken from [BB10, BB11].

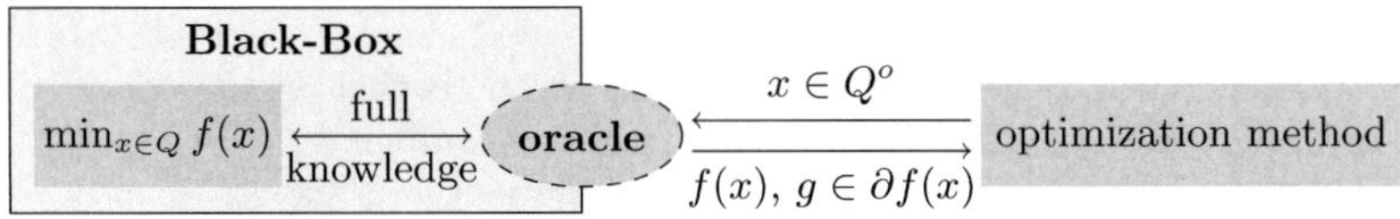

Figure 3.1: *Schematic representation of the Black-Box Optimization idea with a First-Order oracle: in order to have the entities in this illustration well-defined, we assume that the feasible set $Q \subset \mathbb{R}^n$ is closed and convex and that the objective function $f : Q \to \mathbb{R}$ is convex.*

3.1 Black-Box Optimization

Let us start by introducing the Black-Box Optimization concept, which is due to Nemirovski and Yudin [NY83].

In this framework, the optimizer knows in advance several characteristics of the problem she wants to solve. For instance, she might know that her problem is convex, with a differentiable objective, and she might also know the Lipschitz constant of the gradient, and so on. Based on this information, the optimizer chooses an optimization method to solve the particular instance of her problem.

According to the Black-Box Optimization setting, this method does not have access to the problem instance it is dealing with as a whole. Metaphorically speaking, the optimizer hides the full problem instance in a **black box** from the method she has chosen. Importantly, the method is not able to open this box. However, there is an **oracle** that is attached to this black box. An oracle is an entity (a piece of code, for instance), that takes as input a test point and returns some relevant information on the instance, such as the value of the objective function or a subgradient at that point; see Figure 3.1 for a schematic representation. The optimization method calls this oracle and uses its output to construct an approximate solution to the problem instance it is confronted with. Most importantly, the information that the oracle gives should be local, that is, if we modify the problem instance outside of a certain neighborhood of the test point, the oracle answer should not change.

Hiding the full format of the problem instance in a black box has several advantages. In fact, this setting has been created to formalize the complexity study of some families of methods on some classes of problems. For instance, we can define the oracle outputs as nastily as possible, which has resulted in many negative complexity results; a classical account on this topic can be found in [NY83]. Second, using the oracle to outsource the computation of the relevant problem information (objective function value, subgradients, and so on) simplifies the complexity analysis of the optimization method significantly. And last but not least, we prohibit implicitly any assumption on the structure of the problem, which results in optimization methods that are broadly applicable.

3.2 Dual Averaging schemes

Dual Averaging schemes [Nes09] carry the underlying idea of Black-Box Optimization to the extremes. In the analysis of these schemes, we disregard the objective function and give a convergence result that considers exclusively the oracle returns. Clearly, when disregarding the objective function, we cannot make any statement on the Optimality Gap (2.3). We thus need to define at first an alternative quantity which we would like to keep under control by Dual Averaging schemes. In the second part of this section, we introduce Dual Averaging methods and study their analytical complexity. We describe by $Q \subset \mathbb{R}^n$ the set of alternatives and suppose that this set is closed and convex. We assume that we have at our disposal an oracle $\mathcal{G}$. For input $x \in Q$, this oracle returns a vector $g = \mathcal{G}(x) \in \mathbb{R}^n$, which we interpret as loss vector associated to the alternative x. We define the corresponding loss as $\langle g, x \rangle$, meaning that g represents a dual function through the scalar product. Assume now that we play the following iterative game. For $t \in \mathbb{N}_0 := \mathbb{N} \cup \{0\}$, we choose an alternative $x_t \in Q$, call the oracle $\mathcal{G}$ with input x_t, observe a loss vector $g_t = \mathcal{G}(x_t) \in \mathbb{R}^n$, and update the choice x_{t+1} of the alternative. After T rounds, we suffer then a total loss of

$$\mathcal{L}_T := \sum_{t=0}^{T-1} \langle g_t, x_t \rangle .$$

Given these loss vectors $(g_t)_{t=0}^{T-1}$, we would choose by hindsight as alternative

$$x^* \in \arg\min_{x \in Q} \left\{ \sum_{t=0}^{T-1} \langle g_t, x \rangle \right\}$$

in order to have minimal total loss, which we denote by $\mathcal{L}_T^* := \sum_{t=0}^{T-1} \langle g_t, x^* \rangle$. Clearly, the value of the optimal hindsight decision x^* depends on the observed loss vectors $(g_t)_{t=0}^{T-1}$. As these vectors are not known in advance, there is no chance to determine x^* at the beginning. However, we can use the quantity $\mathcal{L}_T^*$ as reference value for the total loss $\mathcal{L}_T$ that we suffer by our online choice of alternatives $(x_t)_{t=0}^{T-1}$. The difference of these two quantities is called regret.

Definition 3.1 (Regret) *Given the vectors $(x_t)_{t=0}^{T-1} \subset Q$ and $(g_t)_{t=0}^{T-1} \subset \mathbb{R}^n$ as above, we define the regret at $T \in \mathbb{N}$ as*

$$\mathcal{R}_T := \sum_{t=0}^{T-1} \langle g_t, x_t \rangle - \min_{x \in Q} \left\{ \sum_{t=0}^{T-1} \langle g_t, x \rangle \right\} = \max_{x \in Q} \left\{ \sum_{t=0}^{T-1} \langle g_t, x_t - x \rangle \right\}.$$

In principle, the regret can take non-positive values, for instance when x_t and g_t satisfy $\langle g_t, x_t \rangle = \min_{x \in Q} \langle g_t, x \rangle$ for any $0 \le t \le T - 1$.

The perception of the losses might be time-dependent. For instance, a loss that we suffer today might be much more painful or costly than a future loss. Or, vice versa, the latter we observe a loss, the more valuable it can be as it is more up-to-date. A simple way of including this kind of considerations in the evaluation is to multiply the losses by time-dependent weights. We refer to the resulting quantity as weighted regret.

Definition 3.2 (Weighted regret) *Let the assumptions of Definition 3.1 hold and choose positive weights $(\gamma_t)_{t=0}^{T-1} \subset \mathbb{R}$. We define the weighted regret at $T \in \mathbb{N}$ as*

$$\mathcal{W}_T := \frac{1}{\sum_{k=0}^{T-1} \gamma_k} \max_{x \in Q} \left\{ \sum_{t=0}^{T-1} \gamma_t \langle g_t, x_t - x \rangle \right\}.$$

Now, the following question arises naturally: is there a way to update the alternatives $(x_t)_{t=0}^{T-1}$ such that the weighted regret $\mathcal{W}_T$ at $T \in \mathbb{N}$ can be bounded from above by a quantity that goes to zero when T goes to infinity? And related to this question: how fast does this quantity go to zero? These questions have been answered in [NY83, Nes09] in full generality, whereas they are discussed in [FS97, AHK05] for a particular setting. We display here the version of Nesterov [Nes09], because it gives the most complete answers to these questions. In Chapter 5, we show that the results in [FS97] can be indeed rediscovered in Nesterov's framework [Nes09].

Let $T \in \mathbb{N}$ and $0 \le t \le T - 1$. In [Nes09], all loss vectors $(g_k)_{k=0}^{t}$ are accumulated in a weighted dual variable $s_t = -\sum_{k=0}^{t-1} \gamma_k g_k$, where $\gamma_k > 0$, $k = 0, \ldots, t - 1$, are some positive weights. To be compatible with the terminology in Convex Optimization, we refer to the weights γ_k as **stepsizes** from now on. The dual variable s_t is then projected (or mirrored) back on the feasible set Q using a **parametrized mirror-operator**. This operator requires a distance-generating function $d : Q \to \mathbb{R}$ and is defined as

$$\pi_{Q,\beta} : \mathbb{R}^n \to Q^o : s \mapsto \arg\min_{x \in Q} \left\{ -\langle s, x \rangle + \beta d(x) \right\}, \tag{3.1}$$

where $\beta > 0$. Recall from Remark 2.2 that this parametrized mirror-operator is well-defined and that its unique minimizer can be written in a closed

---

Algorithm 3.1 Dual Averaging methods [Nes09]

1: Fix $T \in \mathbb{N}$, which corresponds to the total number of iterations.
2: Set $s_0 = 0$.
3: Select positive step-sizes $(\gamma_t)_{t=0}^{T-1}$ and a non-decreasing sequence $(\beta_t)_{t=0}^{T}$ of positive projection parameters.
4: Set $x_0 = c(d) \in Q^o$.
5: **for** $0 \leq t \leq T - 1$ **do**
6:     Call the oracle $\mathcal{G}$ to get an element $g_t = \mathcal{G}(x_t) \in \mathbb{R}^n$.
7:     Set $s_{t+1} = s_t - \gamma_t g_t$.
8:     Compute $x_{t+1} := \pi_{Q,\beta_{t+1}}(s_{t+1}) \in Q^o$.
9: **end for**

---

form due to Assumption 2.1. We just explained the construction of a single iteration of Dual Averaging schemes. For a full description of these methods, we refer to Algorithm 3.1.

We assume that the sequences $(x_t)_{t=0}^{T}$, $(g_t)_{t=0}^{T-1}$, $(\lambda_t)_{t=0}^{T-1}$, and $(\beta_t)_{t=0}^{T}$ are in accordance to the update rules defined in Algorithm 3.1. Nesterov [Nes09] proved the following upper bound on the (restricted) weighted regret.

Theorem 3.1 (Theorem 1 in [Nes09]) *Let* $\Gamma = \sum_{t=0}^{T-1} \gamma_t$ *and* $\Omega_d \geq 0$. *Then,*

$$\frac{1}{\Gamma} \max_{x \in Q : d(x) \leq \Omega_d} \left\{ \sum_{t=0}^{T-1} \gamma_t \langle g_t, x_t - x \rangle \right\} \leq \frac{1}{\Gamma} \left(\beta_T \Omega_d + \frac{1}{2} \sum_{t=0}^{T-1} \frac{\gamma_t^2}{\beta_t} ||g_t||_*^2 \right). \quad (3.2)$$

Let us quickly make a comment on the constant Ω_d in the above theorem. Recall that the set Q is not necessarily bounded. However, there might be no maximizer of a linear function over an unbounded non-empty set. We invoke the additional restriction $d(x) \leq \Omega_d$ to bound the problem in the left-hand side of (3.2) artificially. The constants Ω_d for the simplex and the matrix simplex setup are presented in Example 2.18.

We observe that the right-hand side in (3.2) does not need to converge necessarily. The convergence or divergence of this quantity is determined by the choice of the sequences $(\gamma_t)_{t=0}^{T-1}$ and $(\beta_t)_{t=0}^{T}$. Let $\Omega_d > 0$ and assume that the norm of the vectors g_t is uniformly bounded, that is, there exists a constant M such that $||g_t||_* \leq M$ for any $0 \leq t \leq T - 1$. The most basic choice for the β_t's corresponds to setting all equal to 1. Nesterov [Nes09] observed that in this situation the right-hand side in (3.2) is guaranteed to converge to zero if and only if $\sum_{k=0}^{t} \gamma_k$ diverges and γ_t converges to zero as t goes to infinity. The latter condition says that the weight of the losses is decreasing over time. However, the common sense dictates that new losses should be treated with more consideration than old losses as they are likely to contain more relevant information.

If the time horizon T is finite and known in advance, the requirement $\gamma_t \to 0$ is not necessary any more. In particular, we can take $\gamma_t = \gamma$ as constant for any $0 \le t \le T - 1$. Minimizing the right-hand side in (3.2) with respect to γ, we obtain the following optimal constant step-size policy:

$$\gamma^* = \frac{1}{M} \sqrt{\frac{2\Omega_d}{T}}.$$

With these parameter choices, Inequality (3.2) in Theorem 3.1 can be rewritten as:

$$\frac{1}{T} \max_{x \in Q} \left\{ \sum_{t=0}^{T-1} \langle g_t, x_t - x \rangle : d(x) \le \Omega_d \right\} \le M \sqrt{\frac{2\Omega_d}{T}}. \tag{3.3}$$

That is, for any $\epsilon > 0$, we are guaranteed that the left hand-side in (3.3) is bounded from above by ϵ after at most

$$T = \left\lceil 2\Omega_d M^2 / \epsilon^2 \right\rceil$$

iterations of Algorithm 3.1.

Nesterov derived in his paper [Nes09] a technique that allows to choose non-decreasing step-sizes while still ensuring that the right-hand side in (3.2) converges to zero. For instance, we can choose constant step-sizes $\gamma_t = \bar{\gamma}$ for any $0 \le t \le T - 1$, where $\bar{\gamma} > 0$, and set

$$\beta_0 = 1 \qquad \text{and} \qquad \beta_{t+1} = \sum_{k=0}^{t} 1/\beta_k \qquad \forall \, 0 \le t \le T - 1. \tag{3.4}$$

For these parameters, we can rewrite Inequality (3.2) in Theorem 3.1 as:

$$\frac{1}{T} \max_{x \in Q} \left\{ \sum_{t=0}^{T-1} \langle g_t, x_t - x \rangle : d(x) \le \Omega_d \right\} \le \frac{\beta_T}{T} \left(\frac{\Omega_d}{\bar{\gamma}} + \frac{M^2 \bar{\gamma}}{2} \right). \tag{3.5}$$

The right-hand side of the above inequality is minimized by

$$\bar{\gamma}^* = \frac{\sqrt{2\Omega_d}}{M}.$$

We observe that $\bar{\gamma}^*$ does not depend on T, whereas γ^* does. Consider now the following lemma.

Lemma 3.1 (Lemma 3 in [Nes09]) *With $(\beta_t)_{t=0}^T$ defined as in (3.4), we have for any $T \geq 1$:*

$$\sqrt{2T-1} \leq \beta_T \leq \frac{1}{1+\sqrt{3}} + \sqrt{2T-1}.$$

Using $\bar{\gamma}^*$ at the place of $\bar{\gamma}$ in (3.5) and applying the above lemma, we obtain:

$$\frac{1}{T} \max_{x \in Q : d(x) \leq \Omega_d} \left\{ \sum_{t=0}^{T-1} \langle g_t, x_t - x \rangle \right\} \leq M\sqrt{2\Omega_d} \left(\frac{1}{(1+\sqrt{3})\,T} + \sqrt{\frac{2}{T} - \frac{1}{T^2}} \right).$$

The right-hand side of the above inequality converges to zero with the order $\mathcal{O}(1/\sqrt{T})$. In order to bound the left-hand side of the above inequality by $\epsilon > 0$, we thus need to perform $\mathcal{O}(1/\epsilon^2)$ iterations of Algorithm 3.1.
In Chapter 5, we discuss an alternative choice for the sequences $(\gamma_t)_{t=0}^{T-1}$ and $(\beta_t)_{t=0}^T$.

3.3 Primal-Dual Subgradient methods

Primal-Dual Subgradient methods are particular instances of Dual Averaging schemes. These methods were developed by Nesterov in [Nes09] and – as we will review in this section – generalize Mirror-Descent schemes [NY83] and the standard Gradient method for unconstrained convex optimization problems. In contrast to Dual Averaging schemes, Primal-Dual Subgradient methods make an assumption on the objective function, namely they require its convexity. This allows us to define the oracle returns as subgradients of the objective function and to derive an upper bound on the optimality gap from (3.2).

3.3.1 Exact subgradients

Primal-Dual Subgradient methods consider problems of the form

$$f^* = \min_{x \in Q} f(x),$$

where $Q \subset \mathbb{R}^n$ is a closed and convex set and the function $f : Q \to \mathbb{R}$ is convex. We assume that a First-Order oracle $\mathcal{G}^{\mathrm{FO}}$ is associated to this problem. This oracle returns the objective function value $f(x)$ and a subgradient $g \in \partial f(x)$ when we call it with $x \in Q^o$. Primal-Dual Subgradient methods are Dual Averaging schemes, where we use $\mathcal{G}^{\mathrm{FO}}$ as oracle. The resulting schemes are presented in Algorithm 3.2.
Let the sequences $(x_t)_{t=0}^T$, $(g_t)_{t=0}^{T-1}$, $(\gamma_t)_{t=0}^{T-1}$, and $(\beta_t)_{t=0}^T$ be given by Algorithm 3.2 and set $\Gamma = \sum_{t=0}^{T-1} \gamma_t$. We denote by x^* an optimal solution to

---

Algorithm 3.2 Primal-Dual Subgradient methods [Nes09]

1: Fix $T \in \mathbb{N}$, which corresponds to the total number of iterations.
2: Set $s_0 = 0$.
3: Select positive step-sizes $(\gamma_t)_{t=0}^{T-1}$ and a non-decreasing sequence $(\beta_t)_{t=0}^{T}$ of positive projection parameters.
4: Set $x_0 = c(d) \in Q^o$.
5: **for** $0 \leq t \leq T-1$ **do**
6:     Call the oracle $\mathcal{G}^{\mathrm{FO}}$ to get a subgradient $g_t = \mathcal{G}^{\mathrm{FO}}(x_t) \in \partial f(x_t)$.
7:     Set $s_{t+1} = s_t - \gamma_t g_t$.
8:     Compute $x_{t+1} := \pi_{Q,\beta_{t+1}}(s_{t+1}) \in Q^o$.
9: **end for**

$\min_{x \in Q} f(x)$ and remember that the definition of the parametrized mirror-operator $\pi_{\beta_t,Q}$ requires a distance-generating function $d : Q \to \mathbb{R}$. We assume now that there exists a constant $\Omega_d > 0$ such that $\Omega_d \geq d(x^*)$. As the oracle returns correspond to subgradients of the function f, the (restricted) weighted regret gives as an upper bound on the optimality gap:

$$
\begin{aligned}
\frac{1}{\Gamma} \max_{x \in Q} & \left\{ \sum_{t=0}^{T-1} \gamma_t \langle g_t, x_t - x \rangle : d(x) \leq \Omega_d \right\} \\
\geq \quad & \frac{1}{\Gamma} \max_{x \in Q} \left\{ \sum_{t=0}^{T-1} \gamma_t \left[f(x_t) - f(x) \right] : d(x) \leq \Omega_d \right\} \\
= \quad & \frac{1}{\Gamma} \sum_{t=0}^{T-1} \gamma_t f(x_t) - \min_{x \in Q} \left\{ f(x) : d(x) \leq \Omega_d \right\} \\
\geq \quad & \min_{0 \leq t \leq T-1} f(x_t) - f^*.
\end{aligned}
$$

Applying Theorem 3.1, we obtain the following convergence result for Algorithm 3.2.

Theorem 3.2 [Nes09] *We have:*

$$
\min_{0 \leq t \leq T-1} f(x_t) - f^* \leq \frac{1}{\Gamma} \left(\beta_T \Omega_d + \frac{1}{2} \sum_{t=0}^{T-1} \frac{\gamma_t^2}{\beta_t} \|g_t\|_*^2 \right).
$$

In a similar way, it can be proven that

$$
f(\bar{x}_T) - f^* \leq \frac{1}{\Gamma} \left(\beta_T \Omega_d + \frac{1}{2} \sum_{t=0}^{T-1} \frac{\gamma_t^2}{\beta_t} \|g_t\|_*^2 \right), \qquad \bar{x}_T := \frac{1}{\Gamma} \sum_{t=0}^{T-1} \gamma_t x_t.
$$

Clearly, the strategies that we discussed in the last section for choosing the step-sizes $(\gamma_t)_{t=0}^{T-1}$ and the sequence $(\beta_t)_{t=0}^{T}$ are applicable to Primal-Dual Subgradients methods, as well. If we select the most basic strategy to choose the β_t's in Algorithm 3.2, that is, $\beta_t = 1$ for any $0 \leq t \leq T$, we rediscover **Mirror-Descent schemes**, which are due to Nemirovski and Yudin [NY83]. These methods have been existing more than twenty years before Primal-Dual Subgradient methods. Note the similarities in the names of the methods: "Primal-Dual" refers to the switches between the primal sequence $(x_t)_{t=0}^{T}$ and the dual sequence $(s_t)_{t=0}^{T}$, "Mirror" to the fact that we mirror (or project) the dual points s_t back to feasible set. Recall that subgradients generalize the concept of gradients to non-differentiable convex functions and that gradients point in the opposite direction of the steepest descent. These observations motivate and relate the terms "Descent" and "Subgradient" in the names of the methods.

We can also show that the most elementary Gradient algorithm is a Primal-Dual Subgradient method if the problem is unconstrained. Let us verify this carefully. We assume that we want to minimize a convex and differentiable function $f : \mathbb{R}^n \to \mathbb{R}$ over $\mathbb{R}^n$. Starting from $x_0 \in \mathbb{R}^n$, the most **elementary Gradient algorithm** generates a sequence $x_{t+1} := x_t - \gamma_t \nabla f(x_t)$, where $\gamma_t > 0$ is an appropriate step-size and $\nabla f(x_t)$ denotes the gradient of f at x_t. Alternatively, the point x_{t+1} can be expressed as $x_{t+1} = x_0 - \sum_{k=0}^{t} \gamma_t \nabla f(x_t)$. Let us now look at iteration t of Algorithm 3.2 and assume that

$$\pi_{Q,\beta_{t+1}} \equiv \pi_{\mathbb{R}^n,1} : \mathbb{R}^n \to \mathbb{R}^n : s \to \arg\max_{x \in \mathbb{R}^n} \left\{ \langle s, x \rangle - d_{\mathrm{euc}}^{x_0}(x) \right\}$$

with the distance-generating function $d_{\mathrm{euc}}^{x_0} : \mathbb{R}^n \to \mathbb{R} : x \mapsto \|x - x_0\|_2^2 / 2$ and with $\beta_{t+1} := 1$. For $s_{t+1} = -\sum_{k=0}^{t} \gamma_k \nabla f(x_k)$, we obtain:

$$
\begin{aligned}
x_{t+1} \quad &:= \quad \arg\max_{x \in \mathbb{R}^n} \left\{ \left\langle -\sum_{k=0}^{t} \gamma_k \nabla f(x_k), x - x_0 \right\rangle - \frac{1}{2} \|x - x_0\|_2^2 \right\} \\
&= \quad x_0 - \sum_{k=0}^{t} \gamma_k \nabla f(x_k),
\end{aligned}
$$

which shows that the most elementary Gradient algorithm is a Mirror-Descent scheme (and thus a Primal-Dual Subgradient method), provided that the feasible set is unconstrained.

3.3.2 Stochastic subgradients

In some situations, the objective function might include a stochastic component. For instance, we might consider the objective function

$$\phi : Q \times \Xi \to \mathbb{R} : (x, \xi) \mapsto \phi(x, \xi),$$

where $Q \subset \mathbb{R}^n$ is convex and closed and $(\Xi, \mathcal{B}, \mathbb{P})$ denotes a Borel probability space. That is, the objective function ϕ depends not only on a deterministic decision vector $x \in Q$, but also on a random element $\xi \in \Xi$. One way to deal with this uncertainty is to consider the expectation of ϕ with respect to ξ, that is, we look at

$$f(x) := \mathbb{E}_{\mathbb{P}}\left[\phi(x, \xi)\right],$$

where we assume that $\phi(x, \cdot)$ is $\mathbb{P}$-integrable for every $x \in Q$. In addition, we suppose that $\phi(\cdot, \xi)$ is convex for any $\xi \in \Xi$, implying that we can apply Primal-Dual Subgradient methods to the stochastic optimization problem

$$\min_{x \in Q}\left\{ f(x) = \mathbb{E}_{\mathbb{P}}\left[f(x, \xi)\right] = \int_{\Xi} \phi(x, \xi) d\mathbb{P}(\xi) \right\}, \tag{3.6}$$

provided that we can easily define an oracle that returns subgradients of f (that is, the computation of subgradients of f should be easy). In most cases, it is however impossible to compute the exact value of f at a given point $x \in Q$ or to derive a subgradient of f at x. Instead of having an exact First-Order oracle, that is, an oracle that returns the exact value of a subgradient g of f at $x \in Q$, we need to content ourselves with a **stochastic First-Order oracle** $\tilde{\mathcal{G}}_N^{\mathrm{FO}}$. When we call the stochastic First-Order oracle $\tilde{\mathcal{G}}_N^{\mathrm{FO}}$ with input point $x \in Q$, it returns a sample approximation

$$\tilde{g} = \tilde{g}(\xi_1, \ldots, \xi_N) := \frac{1}{N} \sum_{i=1}^{N} g_x(\xi_i)$$

of $g \in \partial f(x)$, where $\xi_1, \ldots, \xi_N \in \Xi$ are N independent realizations of the random element ξ and $g_x(\xi_i) \in \partial_x \phi(x, \xi_i)$ denotes a subgradient of $\phi(\cdot, \xi_i)$ at x for any $1 \leq i \leq N$. Interestingly enough, Primal-Dual Subgradient methods also work with this stochastic oracle. The stochastic version of Primal-Dual Subgradient methods is given in Algorithm 3.3. Importantly, as the iterates $(x_t)_{t=1}^{T}$ depend on the sample approximations $(\tilde{g}_t)_{t=0}^{T-1}$, they are themselves realizations of random variables. With a slight abuse of notation, we write x_t for both the random variable and its realization. The meaning should be clear from the context.

Denote by x^* an optimal solution to (3.6). Choose $T \in \mathbb{N}$ and assume that the sequences $(x_t)_{t=0}^{T}$, $(\tilde{g}_t)_{t=0}^{T-1}$, $(\lambda_t)_{t=0}^{T-1}$, and $(\beta_t)_{t=0}^{T}$ are given by Algorithm 3.3. We fix $\Omega_d \geq d(x^*)$ and $\Gamma = \sum_{t=0}^{T-1} \gamma_t$. Algorithm 3.3 is a Dual Averaging scheme, implying that we obtain by Theorem 3.1:

$$\frac{1}{\Gamma} \max_{x \in Q}\left\{ \sum_{t=0}^{T-1} \gamma_t \langle \tilde{g}_t, x_t - x \rangle : d(x) \leq \Omega_d \right\} \leq \frac{1}{\Gamma}\left(\beta_T \Omega_d + \frac{1}{2} \sum_{t=0}^{T-1} \frac{\gamma_t^2}{\beta_t} \|\tilde{g}_t\|_*^2 \right).$$

Algorithm 3.3 Stochastic Primal-Dual Subgradient methods [Nes09]

1: Fix $T \in \mathbb{N}$, which corresponds to the total number of iterations.
2: Set $s_0 = 0$ and choose $N \geq 1$.
3: Select positive step-sizes $(\gamma_t)_{t=0}^{T-1}$ and a non-decreasing sequence $(\beta_t)_{t=0}^{T}$ of positive projection parameters.
4: Set $x_0 = c(d) \in Q^o$.
5: **for** $0 \leq t \leq T - 1$ **do**
6:     Call the stochastic oracle $\tilde{\mathcal{G}}_N^{\mathrm{FO}}$ to get an approximation $\tilde{g}_t = \tilde{\mathcal{G}}_N^{\mathrm{FO}}(x_t)$ of $g_t \in \partial f(x_t)$.
7:     Set $s_{t+1} = s_t - \gamma_t \tilde{g}_t$.
8:     Compute $x_{t+1} := \pi_{Q,\beta_{t+1}}(s_{t+1}) \in Q^o$.
9: **end for**

On the other hand, we have:

$$
\frac{1}{\Gamma} \max_{x \in Q} \left\{ \sum_{t=0}^{T-1} \gamma_t \langle \tilde{g}_t, x_t - x \rangle : d(x) \leq \Omega_d \right\}
$$

$$
\geq \frac{1}{\Gamma} \sum_{t=0}^{T-1} \gamma_t \langle \tilde{g}_t, x_t - x^* \rangle
$$

$$
= \frac{1}{\Gamma} \sum_{t=0}^{T-1} \gamma_t \left\langle \frac{1}{N} \sum_{i=1}^{N} g_{x_t}(\xi_{t,i}), x_t - x^* \right\rangle
$$

$$
= \frac{1}{\Gamma} \sum_{t=0}^{T-1} \frac{\gamma_t}{N} \sum_{i=1}^{N} \langle g_{x_t}(\xi_{t,i}), x_t - x^* \rangle
$$

$$
\geq \frac{1}{\Gamma} \sum_{t=0}^{T-1} \frac{\gamma_t}{N} \sum_{i=1}^{N} [\phi(x_t, \xi_{t,i}) - \phi(x^*, \xi_{t,i})],
$$

where the concluding inequality holds as $g_{x_t}(\xi_{t,i})$ is a subgradient of $\phi(\cdot, \xi_{t,i})$ at x_t for any $0 \leq t \leq T - 1$ and any $1 \leq i \leq N$. Thus,

$$
\frac{1}{\Gamma} \left(\beta_T \Omega_d + \frac{1}{2} \sum_{t=0}^{T-1} \frac{\gamma_t^2}{\beta_t} \|\tilde{g}_t\|_*^2 \right)
$$

$$
\geq \frac{1}{\Gamma} \sum_{t=0}^{T-1} \frac{\gamma_t}{N} \sum_{i=1}^{N} [\phi(x_t, \xi_{t,i}) - \phi(x^*, \xi_{t,i})]. \tag{3.7}
$$

Let us assume that there exists a constant $\tilde{M}$ such that

$$
\mathbb{E}_{\mathbb{P}^N}\left[\|\tilde{g}\|_*^2 \right] \leq \tilde{M}^2
$$

for any output $\tilde{g}$ of the oracle $\tilde{\mathcal{G}}_N^{\mathrm{FO}}$. In addition, we observe that x_t and $\xi_{t,1}, \ldots, \xi_{t,N}$ are independent random variables, as x_t depends only on the random variables $\xi_{k,i}$, where $0 \leq k \leq t-1$ and $1 \leq i \leq N$. The independence of these random variables allows us to write $\mathbb{E}_{\mathbb{P}NT}\,\phi(x_t, \xi_{t,i})$ as $\mathbb{E}_{\mathbb{P}NT}\,f(x_t)$ for any $0 \leq t \leq T-1$. Taking expectations on both sides of Inequality (3.7), we thus obtain:

$$
\begin{aligned}
\frac{1}{\Gamma} &\left(\beta_T \Omega_d + \frac{\tilde{M}^2}{2} \sum_{t=0}^{T-1} \frac{\gamma_t^2}{\beta_t} \right) \\
&\geq \quad \frac{1}{\Gamma} \left(\beta_T \Omega_d + \frac{1}{2} \sum_{t=0}^{T-1} \frac{\gamma_t^2}{\beta_t} \mathbb{E}_{\mathbb{P}N} \left[\|\tilde{g}_t\|_*^2 \right] \right) \\
&= \quad \frac{1}{\Gamma} \mathbb{E}_{\mathbb{P}NT} \left[\beta_T \Omega_d + \frac{1}{2} \sum_{t=0}^{T-1} \frac{\gamma_t^2}{\beta_t} \|\tilde{g}_t\|_*^2 \right] \\
&\geq \quad \frac{1}{\Gamma} \mathbb{E}_{\mathbb{P}NT} \left[\sum_{t=0}^{T-1} \frac{\gamma_t}{N} \sum_{i=1}^{N} [\phi(x_t, \xi_{t,i}) - \phi(x^*, \xi_{t,i})] \right] \\
&= \quad \frac{1}{\Gamma} \sum_{t=0}^{T-1} \frac{\gamma_t}{N} \sum_{i=1}^{N} \left(\mathbb{E}_{\mathbb{P}NT} \left[\phi(x_t, \xi_{t,i}) \right] - \mathbb{E}_{\mathbb{P}NT} \left[\phi(x^*, \xi_{t,i}) \right] \right) \\
&= \quad \frac{1}{\Gamma} \sum_{t=0}^{T-1} \frac{\gamma_t}{N} \sum_{i=1}^{N} \left(\mathbb{E}_{\mathbb{P}NT} \left[f(x_t) \right] - f(x^*) \right) \\
&= \quad \frac{1}{\Gamma} \sum_{t=0}^{T-1} \gamma_t \left(\mathbb{E}_{\mathbb{P}NT} \left[f(x_t) \right] - f(x^*) \right).
\end{aligned}
$$

We have verified the following result.

Theorem 3.3 [Nes09] *It holds that:*

$$
\mathbb{E}_{\mathbb{P}NT} \left[\min_{0 \leq t \leq T-1} f(x_t) \right] - f(x^*) \leq \frac{1}{\Gamma} \left(\beta_T \Omega_d + \frac{\tilde{M}^2}{2} \sum_{t=0}^{T-1} \frac{\gamma_t^2}{\beta_t} \right).
$$

Using the convexity of f, we can also show:

$$
\mathbb{E}_{\mathbb{P}NT} \left[f \left(\frac{1}{\Gamma} \sum_{t=0}^{T-1} \gamma_t x_t \right) \right] - f(x^*) \leq \frac{1}{\Gamma} \left(\beta_T \Omega_d + \frac{\tilde{M}^2}{2} \sum_{t=0}^{T-1} \frac{\gamma_t^2}{\beta_t} \right).
$$

3.3.3 How fast can Subgradient methods be?

The Black-Box Optimization framework has been developed to formalize the complexity analysis of some optimization algorithms and has resulted in many lower complexity bounds. Of particular interest to us is the following theorem, which concerns Subgradient methods and shows that these schemes cannot convergence faster than with the order $\Omega(1/\sqrt{T})$. In particular, this result ensures the optimality of the Algorithms 3.2 and 3.3.

Theorem 3.4 (Theorem 3.2.1 in [Nes04]) *Consider the unconstrained minimization problem* $\min f(x)$, *where* $f : \mathbb{R}^n \to \mathbb{R}$ *is a convex function, and let* $x_0 \in \mathbb{R}^n$. *Assume that there exists a minimizer* x^* *of* f, *that* x^* *belongs to* $B := \{x \in \mathbb{R}^n : \|x - x_0\|_2 \leq R\}$ *for an* $R > 0$, *and that* f *is Lipschitz continuous on* B *with Lipschitz constant* M. *We have at our disposal a First-Order oracle* $x \mapsto (f(x), g) \in \mathbb{R} \times \partial f(x)$. *If the optimization method constructs a sequence of points* $(x_t)_{t \geq 0}$ *such that* $x_t \in x_0 + span\{g_0, \ldots, g_{t-1}\}$ *for any* $t \geq 1$, *then:*

$$f(x_t) - f(x^*) \geq \frac{MR}{2(1 + \sqrt{t+1})} \qquad \forall \, 0 \leq t < n.$$

3.4 Optimal First-Order methods

In this section, we consider optimization problems of the form

$$f^* = \min_{x \in Q} f(x),$$

where Q is a closed and convex subset of $\mathbb{R}^n$ and $f : \mathbb{R}^n \to \mathbb{R}$ is a function that is convex on Q and that belongs to $C^1_{L(Q)}$. By Definition (2.19), the constant $L = L(Q) > 0$ satisfies the inequality:

$$\|\nabla f(x) - \nabla f(y)\|_* \leq L \|x - y\| \qquad \forall \, x, y \in Q. \tag{3.8}$$

Nesterov [Nes04, Nes05] designed a method for this problem class; see Equations (5.6) in [Nes05]. This method has a convergence rate of $\mathcal{O}\left(L/T^2\right)$, which outperforms the rate of convergence for Subgradient methods by two orders of magnitude.

At every step of Nesterov's method [Nes04, Nes05], the Lipschitz constant L is used to update the iterates. However, the constant L is a global parameter of the function f, as L needs to satisfy Condition (3.8) on the whole set Q. In this section, we introduce a refined version of Nesterov's algorithm, where we replace the global parameter L by local estimates. Numerical results in Chapter 11 show that this replacement yields to methods with a better performance in practice.

Let $d : Q \to \mathbb{R}$ be a distance-generating function. Recall from Section 2.5 that we write $V_z(x) = d(x) - d(z) - \langle d'(z), x - z \rangle$ for the Bregman distance of

$x \in Q$ with respect to $z \in Q^o$. Nesterov's method requires a **prox-mapping**, that is, a mapping of the form:

$$\text{Prox}_{Q,z} : \mathbb{R}^n \to Q^o : s \mapsto \arg\min_{x \in Q} \left\{ \langle s, x - z \rangle + V_z(x) \right\}, \qquad z \in Q^o. \quad (3.9)$$

Given $s \in \mathbb{R}^n$ and $z \in Q^o$, the prox-mapping $\text{Prox}_{Q,z}(s)$ can be rewritten as

$$\text{Prox}_{Q,z}(s) = \arg\min_{x \in Q} \left\{ \langle s - d'(z), x \rangle + d(x) \right\},$$

which shows that the value $\text{Prox}_{Q,z}(s)$ is well-defined, that it belongs to Q^o (by Remark 2.2), and that it can be written in a closed form (due to Assumption 2.1).

We consider Algorithm 3.4 and let $T \in \mathbb{N}_0$. Provided that we set $L_t = L$ for any $0 \le t \le T$ in Algorithm 3.4, this scheme coincides with Nesterov's method [Nes04, Nes05].

Let us now study the analytical complexity of the modified method. We assume that the sequences $(x_t)_{t=0}^{T+1}$, $(u_t)_{t=0}^{T+1}$, $(z_t)_{t=0}^{T}$, $(\hat{x}_t)_{t=1}^{T+1}$, $(\gamma_t)_{t=0}^{T+1}$, $(\Gamma_t)_{t=0}^{T+1}$, $(\tau_t)_{t=0}^{T}$, and $(L_t)_{t=0}^{T}$ are generated by Algorithm 3.4.

Given $0 \le t \le T$, we say that **Inequality ($\mathcal{I}_t$)** holds if

$$\Gamma_t f(u_t) + \sum_{k=0}^{t-1} (L_{k+1} - L_k) \left(d(z_{k+1}) - \frac{1}{2} \|z_k - \hat{x}_{k+1}\|^2 \right) \le \psi_t, \qquad (\mathcal{I}_t)$$

where

$$\psi_t := \min_{x \in Q} \left\{ \sum_{k=0}^{t} \gamma_k \left(f(x_k) + \langle \nabla f(x_k), x - x_k \rangle \right) + L_t d(x) \right\}.$$

As the proof of the following result is rather long and technical, we give it in the Appendix B.1.

Theorem 3.5 *Inequality ($\mathcal{I}_t$) holds for any $0 \le t \le T$.*

As usual, the element $x^* \in Q$ represents an optimal solution to the optimization problem $f^* = \min_{x \in Q} f(x)$.

Theorem 3.6 *For any $T \in \mathbb{N}_0$, we have:*

$$f(u_T) - f^* \le \frac{1}{\Gamma_T} \left[L_T d(x^*) + \sum_{t=0}^{T-1} (L_t - L_{t+1}) \left(d(z_{t+1}) - \frac{1}{2} \|z_t - \hat{x}_{t+1}\|^2 \right) \right].$$

Algorithm 3.4 Optimal First-Order method with adaptive L-estimation

---

1: Choose $T \in \mathbb{N}_0$.
2: Choose $(\gamma_t)_{t=0}^{T+1}$ with $\gamma_0 \in (0,1]$, $\gamma_t \geq 0$, and $\gamma_t^2 \leq \Gamma_t := \sum_{k=0}^{t} \gamma_k$ for any $0 \leq t \leq T+1$.
3: Set $L_0 = L$ and $x_0 = c(d)$.
4: Compute $u_0 := \arg\min_{x \in Q} \{\gamma_0 \left(f(x_0) + \langle \nabla f(x_0), x - x_0 \rangle\right) + L_0 d(x)\}$.
5: Set $z_0 = u_0$, $\tau_0 = \gamma_1/\Gamma_1$, and $x_1 = \tau_0 z_0 + (1 - \tau_0)u_0 = z_0$.
6: Define $\hat{x}_1 := \mathrm{Prox}_{Q,z}\left(\gamma_1 \nabla f(x_1)/L_0\right)$.
7: Set $u_1 = \tau_0 \hat{x}_1 + (1 - \tau_0)u_0$.
8: **for** $1 \leq t \leq T$ **do**
9:     Choose $0 < L_t \leq L$ such that:

$$f(u_t) \leq f(x_t) + \langle \nabla f(x_t), u_t - x_t \rangle + \frac{L_t}{2} \|u_t - x_t\|^2. \qquad (3.10)$$

10:     Set $z_t = \arg\min_{x \in Q} \left\{\sum_{k=0}^{t} \gamma_k \left(f(x_k) + \langle \nabla f(x_k), x - x_k \rangle\right) + L_t d(x)\right\}$.
11:     Set $\tau_t = \gamma_{t+1}/\Gamma_{t+1}$ and $x_{t+1} = \tau_t z_t + (1 - \tau_t)u_t$.
12:     Compute $\hat{x}_{t+1} := \mathrm{Prox}_{Q,z_t}\left(\gamma_{t+1} \nabla f(x_{t+1})/L_t\right)$.
13:     Set $u_{t+1} = \tau_t \hat{x}_{t+1} + (1 - \tau_t)u_t$.
14: **end for**

Proof: Let $0 \leq t \leq T$. The convexity of the function f and the definition of Γ_t imply:

$$
\begin{aligned}
\psi_t &:= \min_{x \in Q} \left\{ L_t d(x) + \sum_{k=0}^{t} \gamma_k \left(f(x_k) + \langle \nabla f(x_k), x - x_k \rangle\right) \right\} \\
&\leq L_t d(x^*) + \sum_{k=0}^{t} \gamma_k \left(f(x_k) + \langle \nabla f(x_k), x^* - x_k \rangle\right) \\
&\leq L_t d(x^*) + \sum_{k=0}^{t} \gamma_k f(x^*) \\
&= L_t d(x^*) + \Gamma_t f(x^*).
\end{aligned}
$$

It remains to combine this inequality with Theorem 3.5. ∎

Nesterov [Nes04, Nes05] suggests to choose the sequence $(\gamma_t)_{t=0}^{T+1}$ as

$$\gamma_t := \frac{t+1}{2} \qquad \forall\, 0 \leq t \leq T+1. \qquad (3.11)$$

Lemma 2 of [Nes05] shows that we have

$$\tau_t = \frac{2}{t+3} \qquad \forall\, 0 \leq t \leq T$$

and

$$\Gamma_t = \frac{(t+1)(t+2)}{4}, \qquad \gamma_t^2 \le \Gamma_t \qquad \forall\, 0 \le t \le T+1$$

for this choice of the γ_t's. With this setting, Theorem 3.6 results in the following corollary.

Corollary 3.2 *Let us choose the sequence* $(\gamma_t)_{t=0}^{T+1}$ *in Algorithm 3.4 as described in (3.11). Then, we have for any* $T \ge 0$:

$$f(u_T) - f^* \;\le\; \frac{4}{(T+1)(T+2)} L_T d(x^*)$$
$$+ \sum_{t=0}^{T-1} \frac{4\,(L_t - L_{t+1})}{(T+1)(T+2)} \left(d(z_{t+1}) - \frac{1}{2} \|z_t - \hat{x}_{t+1}\|^2 \right).$$

$$\blacksquare$$

In the non-adaptive setting, that is, if $L_t = L$ for any $0 \le t \le T$, the inequality of the above corollary takes the form

$$f(u_T) - f^* \le \frac{4}{(T+1)(T+2)} L d(x^*).$$

According to this inequality, we need to perform at most

$$T = \left\lceil 2\sqrt{Ld(x^*)/\epsilon} - 1 \right\rceil$$

iterations of Algorithm 3.4 to find a point $x \in Q$ satisfying $f(x) - f^* \le \epsilon$, where $\epsilon > 0$ denotes the absolute accuracy. This result is in full accordance with Theorem 2 in [Nes05].
In the adaptive setting, we can bound $L_T d(x^*)$ from above by $Ld(x^*)$. Moreover, we have:

$$\frac{4}{(T+1)(T+2)} \sum_{t=0}^{T-1} (L_t - L_{t+1}) \left(d(z_{t+1}) - \frac{1}{2} \|z_t - \hat{x}_{t+1}\|^2 \right)$$
$$\le \frac{20L \sup_{x \in Q} d(x)}{T+2},$$

which holds due to Inequality (2.14). Thus, Algorithm 3.4 needs at most

$$T = \left\lceil 20L \sup_{x \in Q} d(x)/\epsilon - 2 \right\rceil$$

iterations to find a feasible ϵ-solution, provided that $\sup_{x \in Q} d(x)$ is finite. However, we can always switch back to the non-adaptive setting. That is, we can choose a constant $\mathcal{O}(1)$ and install the following supplementary rule for the choice of the parameters L_t. As soon as

$$\sum_{k=0}^{i-1} (L_k - L_{k+1}) \left(d(z_{k+1}) - \frac{1}{2} \|z_k - \hat{x}_{k+1}\|^2 \right) \geq \mathcal{O}(1) L d(x^*) \qquad (3.12)$$

for some $i \geq 1$, we set $L_t = L$ for all $t > i$. In other words, we switch back to the non-adaptive setting after we have reached a certain threshold defined by $\mathcal{O}(1)$.

Algorithm 3.4 equipped with this additional switch-back rule has - in $\mathcal{O}$-notation – the same worst-case analytical complexity bound as the original method presented in [Nes04, Nes05]. As we will observe in Chapter 11, we can significantly reduce the running time of the method in practice by replacing the global constant L by smaller local estimates $L_t \leq L$ of L.

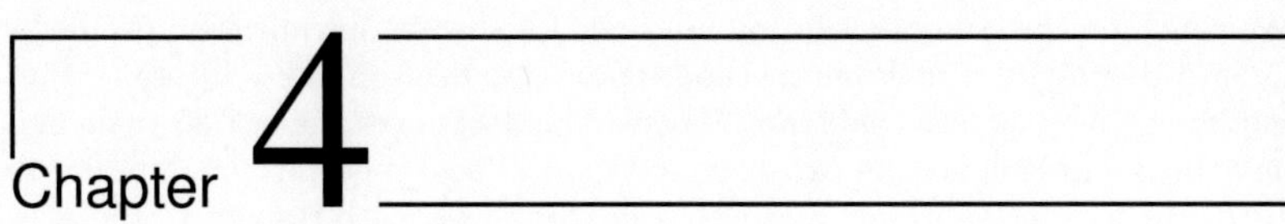

Chapter 4

Solution methods in Structural Optimization

THE key feature of Black-Box Optimization is that the exact form (or the structure) of the problem is hidden from the solution method. However, in order to choose an optimization method, the optimizer needs to verify several characteristics (such as convexity, for instance) of the problem. In particular, she knows the exact form of the problem. Immediately, the following question arises: what is the purpose of putting the problem artificially in a black box? In the last chapter, we gave some reasons that justify this approach. However, the knowledge about the exact problem structure might help us to solve the problem more efficiently. In other words, we might design solution methods that are particularized to a certain problem structure and run much faster on these problem classes than Black-Box optimization schemes. This is the basic motivation of Structural Optimization, where we give the optimization method full access to the problem structure; see [Nes04] for more details.

We study several methods that exploit the particular structure of a problem class in this chapter and show that they significantly outperform traditional Subgradient methods with respect to the analytical complexity. In Section 4.1, we review Smoothing Techniques [Nes05] and Mirror-Prox methods [Nem04a]. These methods are of first-order-type. They can be applied to the same subfamily of non-smooth convex optimization problems and share the same analytical complexity result: about $\mathcal{O}(1/\epsilon)$ iterations of these methods are required to find a feasible ϵ-solution. Smoothing Techniques and Mirror-

Prox methods are probably the most exciting developments in Convex Optimization in the last ten years. Due to these methods, we are nowadays able to solve approximately structured large-scale convex optimization problems whose practical tractability was out of reach ten years ago; see [Ele09, Peñ08] for some examples.

We conclude the study of solution methods for convex optimization problems by introducing Interior-Point methods [Kar84] in Section (4.2). Interior-Point methods have attracted substantial research efforts over the last 30 years and have been implemented in excellent software. These methods are superior to all the other methods that we discuss in this thesis with respect to the analytical complexity. In order to find a feasible ϵ-solution, they require at most $\mathcal{O}(\ln[1/\epsilon])$ iterations, which makes them very well suited for the computation of solutions with a very small approximation error. Under the assumption that a problem satisfies the structural conditions required by Interior-Points methods, these methods are an excellent tool to solve small- and medium-scale problems approximately. However, and in contrast to the other methods presented in this thesis, Interior-Point methods require the computation of second-order information and the resolution of a system of linear equations at every iteration. These requirements hamper the applicability of Interior-Point methods to very large-scale problems. We will discuss this issue in Chapter 6 in more detail.

Contributions and relevant literature: All methods and complexity results presented in this chapter are known. We mainly use the references [JNT08, Nem04a, Nes04, Nes05]. Parts of this review chapter are taken from [BB09, BBN11].

4.1 First-Order methods in Structural Optimization

Traditional Subgradient methods have an analytical complexity of $\mathcal{O}\left(1/\epsilon^2\right)$ for finding a feasible ϵ-solution to convex optimization problems; see Section 3.3. In Section 3.4, we observed that the situation becomes much more favorable if the objective function is not only convex, but also differentiable with a Lipschitz continuous gradient. In this situation, we can approximately solve the problem by performing $\mathcal{O}\left(1/\sqrt{\epsilon}\right)$ iterations of Algorithm 3.4. These two results show a difference of two orders of magnitude with respect to the accuracy. Naturally, the following question arises: is there any First-Order method that fills this gap of the two algorithms for a subfamily of convex optimization problems?

It was a huge breakthrough when Nesterov [Nes05] answered affirmatively this question. He considers a subfamily of convex optimization problems with a very particular structure of non-differentiability. In a first step, he constructs a smooth approximation of the objective function, that is, an auxiliary objective function that is differentiable with a Lipschitz continuous gradient. Then, the resulting auxiliary problem is solved using Algorithm 3.4. In total, this procedure requires $\mathcal{O}\left(1/\epsilon\right)$ iterations of Algorithm 3.4 to derive an approximate solution to the initial problem. Let us now start by introducing Nesterov's procedure.

4.1.1 Smoothing Techniques

We assume that the sets $Q_1 \subset \mathbb{R}^n$ and $Q_2 \subset \mathbb{R}^m$ are both compact and convex. In addition, we endow the spaces $\mathbb{R}^n$ and $\mathbb{R}^m$ with two (maybe different) norms. We denote by $\|\cdot\|_{\mathbb{R}^n}$ and $\|\cdot\|_{\mathbb{R}^m}$ the norm of the spaces $\mathbb{R}^n$ and $\mathbb{R}^m$, respectively. Nesterov considers convex optimization problems of the form:

$$\min_{x \in Q_1} \max_{y \in Q_2} \phi(x,y), \qquad \phi(x,y) := \langle a, x \rangle + \langle \mathcal{A}(x), y \rangle + \langle b, y \rangle, \qquad (4.1)$$

where $a \in \mathbb{R}^n$, $b \in \mathbb{R}^m$, and $\mathcal{A} : \mathbb{R}^n \to \mathbb{R}^m$ is a linear operator. Note that we write $\langle \cdot, \cdot \rangle$ for the Euclidean scalar product in both spaces $\mathbb{R}^n$ and $\mathbb{R}^m$. In principle, the analysis would also work for

$$\tilde{\phi}(x,y) := f_1(x) + \langle \mathcal{A}(x), y \rangle - f_2(y)$$

with convex and smooth functions f_1 and f_2. However, some of the subsequent results would become slightly worse (especially in Section 4.1.2, where we would obtain larger constant factors for some estimates and which would result in smaller step-sizes for the algorithms).

According to the standard MiniMax Theorem in Convex Analysis (see Corollary 37.3.2 in [Roc70]), we have due to the compactness and convexity of the sets Q_1 and Q_2 the following pair of primal-dual convex optimization problems:

$$\min_{x \in Q_1} \left\{ \overline{\phi}(x) := \max_{y \in Q_2} \phi(x,y) \right\} = \max_{y \in Q_2} \left\{ \underline{\phi}(y) := \min_{x \in Q_1} \phi(x,y) \right\}.$$

The operator $\mathcal{A}$ comes with an adjoint operator $\mathcal{A}^* : \mathbb{R}^m \to \mathbb{R}^n$, which is defined by the relation:

$$\langle \mathcal{A}(x), y \rangle = \langle x, \mathcal{A}^*(y) \rangle \qquad \forall\, (x,y) \in \mathbb{R}^n \times \mathbb{R}^m.$$

The analysis of Nesterov's Smoothing Techniques requires an operator norm of $\mathcal{A}$. This norm is constructed as follows.

Definition 4.1 (Operator norm) *The norm of the operator $\mathcal{A} : \mathbb{R}^n \to \mathbb{R}^m$ is defined as:*

$$\|\mathcal{A}\|_{\mathbb{R}^n, \mathbb{R}^m} := \max_{x \in \mathbb{R}^n, y \in \mathbb{R}^m} \left\{ \langle \mathcal{A}(x), y \rangle : \|x\|_{\mathbb{R}^n} = 1, \ \|y\|_{\mathbb{R}^m} = 1 \right\}.$$

Note that the operator norm can be expressed as

$$
\begin{aligned}
\|\mathcal{A}\|_{\mathbb{R}^n, \mathbb{R}^m} &= \max_{x \in \mathbb{R}^n} \left\{ \|\mathcal{A}(x)\|_{\mathbb{R}^m, *} : \ \|x\|_{\mathbb{R}^n} = 1 \right\} \\
&= \max_{y \in \mathbb{R}^m} \left\{ \|\mathcal{A}^*(y)\|_{\mathbb{R}^n, *} : \ \|y\|_{\mathbb{R}^m} = 1 \right\},
\end{aligned}
$$

where $\|\cdot\|_{\mathbb{R}^n, *}$ and $\|\cdot\|_{\mathbb{R}^m, *}$ denote the dual norms of $\|\cdot\|_{\mathbb{R}^n}$ and $\|\cdot\|_{\mathbb{R}^m}$, respectively. The above expression gives rise to the inequalities:

$$\|\mathcal{A}^*(y)\|_{\mathbb{R}^n, *} \leq \|\mathcal{A}\|_{\mathbb{R}^n, \mathbb{R}^m} \|y\|_{\mathbb{R}^m} \ \text{ and } \ \|\mathcal{A}(x)\|_{\mathbb{R}^m, *} \leq \|\mathcal{A}\|_{\mathbb{R}^n, \mathbb{R}^m} \|x\|_{\mathbb{R}^n}, \quad (4.2)$$

where $x \in \mathbb{R}^n$ and $y \in \mathbb{R}^m$, respectively.

We are ready to form a smooth approximation of $\overline{\phi}$ to which we can apply Algorithm 3.4. We choose a distance-generating function $d_{Q_2} : Q_2 \to \mathbb{R}$ and consider the auxiliary function

$$\overline{\phi}_\mu : \mathbb{R}^n \to \mathbb{R} : x \mapsto \max_{y \in Q_2} \left\{ \langle a, x \rangle + \langle \mathcal{A}(x) + b, y \rangle - \mu d_{Q_2}(y) \right\},$$

where $\mu > 0$ is a positive smoothness parameter. This function defines a uniform approximation of $\overline{\phi}$, as

$$\overline{\phi}_\mu(x) \leq \overline{\phi}(x) \leq \overline{\phi}_\mu(x) + \mu \max_{z \in Q_2} d_{Q_2}(z) \qquad \forall \, x \in Q_1; \qquad (4.3)$$

see Inequality (2.7) in [Nes05]. The function $y \mapsto \langle \mathcal{A}(x) + b, y \rangle - \mu d_{Q_2}(y)$ is strongly concave (that is, $y \mapsto -(\langle \mathcal{A}(x) + b, y \rangle - \mu d_{Q_2}(y))$ is strongly convex) for any $x \in Q_1$, as the distance-generating function d_{Q_2} is strongly convex by its definition. Hence, the function $y \mapsto \langle \mathcal{A}(x) + b, y \rangle - \mu d_{Q_2}(y)$ has a unique maximizer on Q_2. We denote this maximizer by $y_*(x)$.

Nesterov showed that $\overline{\phi}_\mu$ is differentiable with a Lipschitz continuous gradient.

Theorem 4.1 (Theorem 1 in [Nes05]) *The function $\overline{\phi}_\mu$ is well-defined, continuously differentiable, and convex on $\mathbb{R}^n$. The gradient of $\overline{\phi}_\mu$ takes the form*

$$\nabla \overline{\phi}_\mu(x) = a + \mathcal{A}^*(y_*(x)),$$

and is Lipschitz continuous with the constant $L_\mu = L_\mu(\mathbb{R}^n) := \|\mathcal{A}\|_{\mathbb{R}^n, \mathbb{R}^m}^2 / \mu$.

Algorithm 4.1 Algorithm 3.4 (with $\gamma_t = (t+1)/2$) applied to Problem (4.4)

1: Choose $T \in \mathbb{N}_0$.

2: Choose a smoothness parameter $\mu > 0$ and a distance-generating function $d_{Q_1} : Q_1 \to \mathbb{R}$.

3: Set $L_0 = L_\mu = \|\mathcal{A}\|^2_{\mathbb{R}^n,\mathbb{R}^m} / \mu$ and $x_0 = c(d_{Q_1})$.

4: Set $u_0 = \arg\min_{x \in Q_1} \left\{ \frac{1}{2} \left(\overline{\phi}_\mu(x_0) + \langle \nabla\overline{\phi}_\mu(x_0), x - x_0 \rangle \right) + L_0 d_{Q_1}(x) \right\}$.

5: Set $z_0 = u_0$, $\tau_0 = \frac{2}{3}$, and $x_1 = \tau_0 z_0 + (1 - \tau_0)u_0 = z_0$.

6: Define $\hat{x}_1 := \mathrm{Prox}_{Q_1,z}\left(\nabla\overline{\phi}_\mu(x_1)/L_0 \right)$.

7: Set $u_1 = \tau_0 \hat{x}_1 + (1 - \tau_0)u_0$.

8: **for** $1 \leq t \leq T$ **do**

9:    Choose $0 < L_t \leq L_\mu$ such that:

$$\overline{\phi}_\mu(u_t) \leq \overline{\phi}_\mu(x_t) + \langle \nabla\overline{\phi}_\mu(x_t), u_t - x_t \rangle + \frac{L_t}{2} \|u_t - x_t\|^2_{\mathbb{R}^n}.$$

10:    Set

$$z_t = \arg\min_{x \in Q_1} \left\{ \sum_{k=0}^{t} \frac{k+1}{2} \left(\overline{\phi}_\mu(x_k) + \langle \nabla\overline{\phi}_\mu(x_k), x - x_k \rangle \right) + L_t d_{Q_1}(x) \right\}.$$

11:    Set $\tau_t = \frac{2}{t+3}$ and $x_{t+1} = \tau_t z_t + (1 - \tau_t)u_t$.

12:    Compute $\hat{x}_{t+1} := \mathrm{Prox}_{Q_1,z_t}\left(\frac{t+2}{2} \nabla\overline{\phi}_\mu(x_{t+1})/L_t \right)$.

13:    Set $u_{t+1} = \tau_t \hat{x}_{t+1} + (1 - \tau_t)u_t$.

14: **end for**

As an immediate consequence, we can apply Algorithm 3.4 to the problem:

$$\min_{x \in Q_1} \overline{\phi}_\mu(x). \tag{4.4}$$

Algorithm 4.1 corresponds to Algorithm 3.4 when we apply this method with step-sizes as described in (3.11) to Problem (4.4).

A slight adaptation of the proof of Theorem 3 in [Nes05] yields to the following result. This slight adaption of the proof is required, as we use Algorithm 3.4 with local estimates for L_μ and not the original method with $L_t = L_\mu$. We need the definitions:

$$D_1 := \max_{x \in Q_1} d_{Q_1}(x) \qquad \text{and} \qquad D_2 := \max_{y \in Q_2} d_{Q_2}(y).$$

Theorem 4.2 *Fix $T \in \mathbb{N}_0$ and assume that the sequences $(x_t)_{t=0}^{T+1}$, $(u_t)_{t=0}^{T+1}$, $(z_t)_{t=0}^{T}$, $(\hat{x}_t)_{t=1}^{T+1}$, and $(L_t)_{t=0}^{T}$ are generated by Algorithm 4.1 for the smoothness parameter*

$q : \mathbb{R} \times \mathbb{R} \to \mathbb{R}$ *such that:*

$$\mu := \frac{2\,\|\mathcal{A}\|_{\mathbb{R}^n,\mathbb{R}^m}}{T+1}\sqrt{\frac{D_1}{D_2}}.$$

For

$$\bar{x} := u_T \in Q_1 \qquad and \qquad \bar{y} := \sum_{t=0}^{T} \frac{2(t+1)}{(T+1)(T+2)} y_*(x_t) \in Q_2,$$

we have:

$$0 \le \overline{\phi}(\bar{x}) - \underline{\phi}(\bar{y}) \le \frac{4\,\|\mathcal{A}\|_{\mathbb{R}^n,\mathbb{R}^m}\sqrt{D_1 D_2}}{T+1} - \frac{4\chi_T}{(T+1)^2},$$

where

$$\chi_T := \sum_{t=0}^{T-1} (L_{t+1} - L_t)\left(d_{Q_1}(z_{t+1}) - \frac{1}{2}\|z_t - \hat{x}_{t+1}\|_{\mathbb{R}^n}^2\right).$$

For the non-adaptive setting, that is, with $L_t = L_\mu$ for any $0 \le t \le T$, we recover Theorem 3 in [Nes05]. In this setting, we need to perform at most

$$T = \left\lceil 4\,\|\mathcal{A}\|_{\mathbb{R}^n,\mathbb{R}^m}\sqrt{D_1 D_2}/\epsilon - 1\right\rceil$$

iterations of Algorithm 4.1 in order to find a tuple $(\bar{x}, \bar{y}) \in Q_1 \times Q_2$ that satisfies $\overline{\phi}(\bar{x}) - \underline{\phi}(\bar{y}) \le \epsilon$, where $\epsilon > 0$.

Proof: Let $T \in \mathbb{N}_0$ and $\mu > 0$. We suppose that the sequences $(x_t)_{t=0}^{T+1}$, $(u_t)_{t=0}^{T+1}$, $(z_t)_{t=0}^{T}$, $(\hat{x}_t)_{t=1}^{T+1}$, and $(L_t)_{t=0}^{T}$ are generated by Algorithm 4.1. Let $\bar{x}$, $\bar{y}$, and χ_T be defined as in the above theorem. In accordance to Theorem 3.5 (that is, we apply Inequality $(\mathcal{I}_T)$) and to the step-size choice (3.11), we have the inequality:

$$\overline{\phi}_\mu(\bar{x}) = \overline{\phi}_\mu(u_T) \le \frac{4(L_T D_1 - \chi_T)}{(T+1)(T+2)} + \min_{x \in Q_1} \frac{2\beta_T(x)}{(T+1)(T+2)}, \tag{4.5}$$

where

$$\beta_T(x) := \sum_{t=0}^{T} (t+1)\left(\overline{\phi}_\mu(x_t) + \langle \nabla\overline{\phi}_\mu(x_t), x - x_t \rangle\right) \qquad \forall\, x \in Q_1.$$

Let $x \in Q_1$. Using Theorem 4.1, we can write:

$$\beta_T(x) \;\; = \;\; \sum_{t=0}^{T} (t+1)\left(\langle a, x \rangle + \langle \mathcal{A}(x) + b, y_*(x_t) \rangle - \mu d_{Q_2}(y_*(x_t))\right)$$

$$\leq \sum_{t=0}^{T}(t+1)\left(\langle a,x\rangle + \langle \mathcal{A}(x)+b, y_*(x_t)\rangle\right)$$

$$= \frac{(T+1)(T+2)}{2}\left(\langle a,x\rangle + \langle \mathcal{A}(x)+b, \bar{y}\rangle\right).$$

The above inequality implies:

$$\min_{x\in Q_1}\beta_T(x) \leq \frac{(T+1)(T+2)}{2}\underline{\phi}(\bar{y}). \tag{4.6}$$

Using that $L_T \leq L_\mu = \|\mathcal{A}\|^2_{\mathbb{R}^n,\mathbb{R}^m}/\mu$ by construction and applying Inequalities (4.5), (4.6), and (4.3), we obtain:

$$
\begin{aligned}
\frac{4(D_1\|\mathcal{A}\|^2_{\mathbb{R}^n,\mathbb{R}^m}/\mu - \chi_T)}{(T+1)^2} &\geq \frac{4(L_T D_1 - \chi_T)}{(T+1)(T+2)} \\
&\geq \overline{\phi}_\mu(\bar{x}) - \underline{\phi}(\bar{y}) \\
&\geq \overline{\phi}(\bar{x}) - \underline{\phi}(\bar{y}) - \mu D_2.
\end{aligned}
\tag{4.7}
$$

It remains to minimize $\frac{4D_1\|\mathcal{A}\|^2_{\mathbb{R}^n,\mathbb{R}^m}}{\mu(T+1)^2} + \mu D_2$ with respect to $\mu > 0$. $\blacksquare$

There exist several strategies to ensure convergence of Algorithm 4.1. First, we may equip Algorithm 4.1 with the switch-back rule described in (3.12). Then, Algorithm 4.1 has – in $\mathcal{O}$-notation – the same worst-case running time as the non-adaptive original method.

On the other hand, we can give the following upper bound for the quantity $(-\chi_T)$ in Theorem 4.2:

$$-\chi_T \leq 5L_\mu D_1 T = \frac{5D_1\|\mathcal{A}\|^2_{\mathbb{R}^n,\mathbb{R}^m}T}{\mu}.$$

This allows us to rewrite (4.7) in the above proof as:

$$\overline{\phi}(\bar{x}) - \underline{\phi}(\bar{y}) \leq \frac{20D_1\|\mathcal{A}\|^2_{\mathbb{R}^n,\mathbb{R}^m}}{(T+1)\mu} + \mu D_2.$$

The right-hand side of the above inequality is minimized by:

$$\mu = 2\|\mathcal{A}\|_{\mathbb{R}^n,\mathbb{R}^m}\sqrt{\frac{5D_1}{(T+1)D_2}},$$

for which we obtain:

$$\overline{\phi}(\bar{x}) - \underline{\phi}(\bar{y}) \leq 4\|\mathcal{A}\|_{\mathbb{R}^n,\mathbb{R}^m}\sqrt{\frac{5D_1 D_2}{T+1}}.$$

Note that this worst-case bound is weaker than the corresponding result for the non-adaptive method.

4.1.2 Mirror-Prox methods

Inspired by Nesterov's work [Nes05], Nemirovski suggested in [Nem04a] an alternative method to solve approximately convex optimization problems satisfying the very specific structural requirements described in (4.1). While Nesterov derives a smooth approximation of the initial problem at first and runs Algorithm 3.4 on the resulting auxiliary problem afterwards, Nemirovski's method is directly applied to the underlying saddle-point problem. Although the algorithmic concepts are completely different, the two methods share the same analytical complexity. Both methods require $\mathcal{O}(\|\mathcal{A}\|_{\mathbb{R}^n,\mathbb{R}^m}/\epsilon)$ iterations to find a feasible ϵ-solution to (4.1).

Together with Saddle-Point Problem (4.1), we can consider the linear operator

$$F : Q_1 \times Q_2 \to \mathbb{R}^n \times \mathbb{R}^m : (x,y) \mapsto \left(\frac{\partial \phi(x,y)}{\partial x}, -\frac{\partial \phi(x,y)}{\partial y} \right),$$

that is, $F(x,y) = (a + \mathcal{A}^*(y), -\mathcal{A}(x) - b)$ for any $(x,y) \in Q_1 \times Q_2$. As for Smoothing Techniques, we endow the set sets Q_1 and Q_2 with two distance-generating functions $d_{Q_1} : Q_1 \to \mathbb{R}$ and $d_{Q_2} : Q_2 \to \mathbb{R}$, respectively. The set $Q_1 \times Q_2$ is a subset of the product space $\mathbb{R}^n \times \mathbb{R}^m$. We equip this space with the norm

$$\|(x,y)\|_{\mathbb{R}^n \times \mathbb{R}^m} := \sqrt{\frac{1}{\Omega_V^2(d_{Q_1})}\|x\|_{\mathbb{R}^n}^2 + \frac{1}{\Omega_V^2(d_{Q_2})}\|y\|_{\mathbb{R}^m}^2}$$

for any $(x,y) \in \mathbb{R}^n \times \mathbb{R}^m$, where $\Omega_V(d_{Q_1})$ and $\Omega_V(d_{Q_2})$ denote the d_{Q_1}- and the d_{Q_2}-diameter of the sets Q_1 and Q_2, respectively; see Definition 2.26. The dual norm takes the form:

$$\|(u,v)\|_{\mathbb{R}^n \times \mathbb{R}^m,*} := \sqrt{\Omega_V^2(d_{Q_1})\|u\|_{\mathbb{R}^n,*}^2 + \Omega_V^2(d_{Q_2})\|v\|_{\mathbb{R}^m,*}^2}$$

for any $(u,v) \in \mathbb{R}^n \times \mathbb{R}^m$. The operator F has the following property.

Lemma 4.1 *The operator F is Lipschitz-continuous with constant*

$$L := \Omega_V(d_1)\Omega_V(d_2)\|\mathcal{A}\|_{\mathbb{R}^n,\mathbb{R}^m},$$

that is,

$$\|F(x,y) - F(x',y')\|_{\mathbb{R}^n \times \mathbb{R}^m,*} \leq L\,\|(x,y) - (x',y')\|_{\mathbb{R}^n \times \mathbb{R}^m}$$

for any $(x,y),(x',y') \in Q_1 \times Q_2$.

Algorithm 4.2 Mirror-Prox methods (adapted from [Nem04a])

1: Choose $T \in \mathbb{N}$.
2: Set $z_0 = (x_0, y_0) = c(d_{Q_1 \times Q_2}) \in Q_1^o \times Q_2^o$.
3: **for** $1 \leq t \leq T$ **do**
4:     Choose $\gamma_t > 0$ such that:

$$\gamma_t \leq \frac{1}{L}. \tag{4.10}$$

5:     Compute $F(z_{t-1}) = F(x_{t-1}, y_{t-1})$.
6:     Set $w_t = \mathrm{Prox}_{Q_1 \times Q_2, z_{t-1}}(\gamma_t F(z_{t-1})) \in Q_1^o \times Q_2^o$.
7:     Compute $F(w_t)$.
8:     Set $z_t = (x_t, y_t) = \mathrm{Prox}_{Q_1 \times Q_2, z_{t-1}}(\gamma_t F(w_t)) \in Q_1^o \times Q_2^o$.
9: **end for**
10: Return $\bar{z} := \left(\sum_{t=1}^{T} \gamma_t \right)^{-1} \sum_{t=1}^{T} \gamma_t w_t$.

Proof: For any $(x, y), (x', y') \in Q_1 \times Q_2$, we have:

$$
\begin{aligned}
\left\| F(x,y) - F(x',y') \right\|_{\mathbb{R}^n \times \mathbb{R}^m, *}^2 & \\
= \; & \Omega_V^2(d_{Q_1}) \left\| \mathcal{A}^*(y - y') \right\|_{\mathbb{R}^n, *}^2 + \Omega_V^2(d_{Q_2}) \left\| \mathcal{A}(x - x') \right\|_{\mathbb{R}^m, *}^2 \\
\leq \; & \Omega_V^2(d_{Q_1}) \left\| \mathcal{A} \right\|_{\mathbb{R}^n, \mathbb{R}^m}^2 \left\| y - y' \right\|_{\mathbb{R}^m}^2 + \Omega_V^2(d_{Q_2}) \left\| \mathcal{A} \right\|_{\mathbb{R}^n, \mathbb{R}^m}^2 \left\| x - x' \right\|_{\mathbb{R}^n}^2 \\
= \; & \Omega_V^2(d_{Q_1}) \Omega_V^2(d_{Q_2}) \left\| \mathcal{A} \right\|_{\mathbb{R}^n, \mathbb{R}^m}^2 \left\| (x - x', y - y') \right\|_{\mathbb{R}^n \times \mathbb{R}^m}^2 ,
\end{aligned}
$$

where the inequality is due to (4.2).   ∎
We consider the function

$$d_{Q_1 \times Q_2} : Q_1 \times Q_2 \to \mathbb{R} : (x,y) \mapsto \frac{1}{\Omega_V^2(d_{Q_1})} d_{Q_1}(x) + \frac{1}{\Omega_V^2(d_{Q_2})} d_{Q_2}(y). \tag{4.8}$$

As proven in Lemma 4.1 in [JNT08], this is a distance-generating function defined on the set $Q_1 \times Q_2$ with the following properties:

$$\Omega_V(d_{Q_1 \times Q_2}) = \sqrt{2} \qquad \text{and} \qquad c(d_{Q_1 \times Q_2}) = (c(d_{Q_1}), c(d_{Q_2})). \tag{4.9}$$

4.1.2.1 Original Mirror-Prox methods

In their original form, which is presented in Algorithm 4.2, Mirror-Prox methods aim at solving Saddle-Point Problem (4.1) when exact values of F are available.

The analytical complexity study of this method requires the following result.

Lemma 4.2 (Lemma 5.3 in [JNT08]) *Let $z \in Q_1^o \times Q_2^o$, $\gamma > 0$, and choose $u, v \in \mathbb{R}^n \times \mathbb{R}^m$. Consider the points*

$$
\begin{aligned}
w &:= \arg \min_{p \in Q_1 \times Q_2} \left\{ \langle \gamma u - d'_{Q_1 \times Q_2}(z), p \rangle + d_{Q_1 \times Q_2}(p) \right\}, \\
z_+ &:= \arg \min_{p \in Q_1 \times Q_2} \left\{ \langle \gamma v - d'_{Q_1 \times Q_2}(z), p \rangle + d_{Q_1 \times Q_2}(p) \right\}.
\end{aligned}
$$

Let $p \in Q_1 \times Q_2$. Then,

$$
\langle \gamma v, w - p \rangle \leq V_z(p) - V_{z_+}(p) - V_z(z_+) + \langle \gamma v, w - z_+ \rangle, \quad (4.11)
$$

where we moreover have:

$$
\begin{aligned}
&-V_z(z_+) + \langle \gamma v, w - z_+ \rangle \\
&\leq \frac{\gamma^2}{2} \|u - v\|_{\mathbb{R}^n \times \mathbb{R}^m, *}^2 - \frac{1}{2} \|w - z\|_{\mathbb{R}^n \times \mathbb{R}^m}^2. \quad (4.12)
\end{aligned}
$$

We display here the lemma taken from [JNT08] and not the initial version of this result presented in [Nem04a], as the more recent version is slightly stronger than its predecessor. Basically, the stronger version allows us to take steps of size at most $1/L$, whereas the step-sizes in [Nem04a] are bounded from above by $1/(\sqrt{2}L)$. As we will observe in Chapter 11, larger step-sizes remarkably improve the the practical behavior of the algorithm.

Using this result in the proof of Theorem 3.2 in [Nem04a], we obtain the following result on the performance of Algorithm 4.2.

Theorem 4.3 (Adapted version of Theorem 3.2 in [Nem04a]) *Let Algorithm 4.2 return $\bar{z} = (\bar{x}, \bar{y}) \in Q_1 \times Q_2$. Then,*

$$
\overline{\phi}(\bar{x}) - \underline{\phi}(\bar{y}) \leq \frac{1}{\sum_{t=1}^{T} \gamma_t}.
$$

Provided that we use constant step-sizes $\gamma_t = 1/L$ for any $1 \leq t \leq T$, the above inequality takes the form:

$$
\overline{\phi}(\bar{x}) - \underline{\phi}(\bar{y}) \leq \frac{\Omega_V(d_{Q_1})\Omega_V(d_{Q_2}) \|A\|_{\mathbb{R}^n, \mathbb{R}^m}}{T}.
$$

In this situation, we need to perform at most

$$
T = \left\lceil \frac{\Omega_V(d_{Q_1})\Omega_V(d_{Q_2}) \|A\|_{\mathbb{R}^n, \mathbb{R}^m}}{\epsilon} \right\rceil
$$

iterations of Algorithm 4.2 to find $(\bar{x}, \bar{y}) \in Q_1 \times Q_2$ with $\overline{\phi}(\bar{x}) - \underline{\phi}(\bar{y}) \leq \epsilon$, where $\epsilon > 0$.

We give here the proof of the above theorem. Based on the proof, we can discuss a strategy to replace the constant L by some local substitutes that we update dynamically.

Proof: Let the sequences $(w_t)_{t=1}^T = ((u_t, v_t))_{t=1}^T$, $(z_t)_{t=0}^T$, and $(\gamma_t)_{t=1}^T$ and the point $\bar{z} = (\bar{x}, \bar{y})$ be given by Algorithm 4.2. Choose $p = (q, r) \in Q_1 \times Q_2$. As ϕ is bilinear, we have:

$$
\begin{aligned}
\sum_{t=1}^T \gamma_t \langle F(w_t), w_t - p \rangle
&= \sum_{t=1}^T \gamma_t \left(\phi(u_t, v_t) - \phi(q, v_t) + \phi(u_t, r) - \phi(u_t, v_t) \right) \\
&= \sum_{t=1}^T \gamma_t \left(\phi(u_t, r) - \phi(q, v_t) \right) \\
&= \sum_{t=1}^T \gamma_t \left(\phi(\bar{x}, r) - \phi(q, \bar{y}) \right) \\
&\geq \sum_{t=1}^T \gamma_t \left(\overline{\phi}(\bar{x}) - \underline{\phi}(\bar{y}) \right),
\end{aligned}
$$

where the inequality is due to the definitions of $\overline{\phi}$ and $\underline{\phi}$. Using Inequality (4.11), Definition (2.26), and Equation (4.9), we can verify the following relations:

$$
\begin{aligned}
\sum_{t=1}^T & \gamma_t \langle F(w_t), w_t - p \rangle \\
&\leq V_{c(d_{Q_1 \times Q_2})}(p) + \sum_{t=1}^T \left(-V_{z_{t-1}}(z_t) + \langle \gamma_t F(w_t), w_t - z_t \rangle \right) \\
&\leq \frac{\Omega_V^2(d_{Q_1 \times Q_2})}{2} + \sum_{t=1}^T \left(-V_{z_{t-1}}(z_t) + \langle \gamma_t F(w_t), w_t - z_t \rangle \right) \\
&= 1 + \sum_{t=1}^T \left(-V_{z_{t-1}}(z_t) + \langle \gamma_t F(w_t), w_t - z_t \rangle \right).
\end{aligned}
$$

We use now Inequality (4.12), the Lipschitz continuity of F, and Condition (4.10) to show that

$$
\sum_{t=1}^T \left(-V_{z_{t-1}}(z_t) + \langle \gamma_t F(w_t), w_t - z_t \rangle \right) \leq 0.
$$

Indeed, we have for any $1 \leq t \leq T$:

$$-V_{z_{t-1}}(z_t) + \langle \gamma_t F(w_t), w_t - z_t \rangle$$

$$\leq \frac{\gamma_t^2}{2} \|F(w_t) - F(z_{t-1})\|^2_{\mathbb{R}^n \times \mathbb{R}^m, *} - \frac{1}{2} \|w_t - z_{t-1}\|^2_{\mathbb{R}^n \times \mathbb{R}^m}$$

$$\leq \frac{L^2 \gamma_t^2}{2} \|w_t - z_{t-1}\|^2_{\mathbb{R}^n \times \mathbb{R}^m} - \frac{1}{2} \|w_t - z_{t-1}\|^2_{\mathbb{R}^n \times \mathbb{R}^m}$$

$$\leq 0.$$

∎

Condition (4.10) appears in the above proof only in the last step, where we use it to ensure the relation $-V_{z_{t-1}}(z_t) + \langle \gamma_t F(w_t), w_t - z_t \rangle \leq 0$ for any $1 \leq t \leq T$. Instead of enforcing (4.10) in Algorithm 4.2, we can proceed as suggested in [Nem04a]: given a point $z_{t-1} \in Q_1 \times Q_2$, we use a computationally more attractive step-size $\tilde{\gamma}_t \geq 1/L$ instead of $\gamma_t \leq 1/L$, compute w_t and z_t, and verify the condition $-V_{z_{t-1}}(z_t) + \langle \tilde{\gamma}_t F(w_t), w_t - z_t \rangle \leq 0$. We refer to [Nem04a] for a possible adjustment strategy of the step-sizes $\tilde{\gamma}_t$. We can view this dynamic step-size strategy as a replacement of Condition (4.10), where we substitute the Lipschitz constant L by local estimates in order to get a less restrictive condition on the step-sizes. Compared to the L-substitution by local estimates in Algorithm 4.1, the dynamic step-size strategy for Mirror-Prox methods has a intrinsic drawback. In Algorithm 4.1, we can validate the current L-replacement before we use it. In Mirror-Prox methods however, the validation of the L-replacement requires the next iterates w_t and z_t, which implies that we may need to adapt the L-replacement in hindsight and to recompute the iterates w_t and z_t. This adaptive process might even require several iterations.

4.1.2.2 Mirror-Prox methods with noisy first-order information

More recently, Juditsky et al. [JNT08] have introduced stochastic Mirror-Prox algorithms, where they allow noisy first-order information. In contrast to the original Mirror-Prox algorithm [Nem04a], they replace the true value of the operator F by an approximate substitute.

More precisely, let us assume that $(\Xi, \mathcal{B}, \mathbb{P})$ is a probability space. The operator $F(z)$, $z \in Q$, is approximated by $\hat{F}_\xi(z)$, where $\hat{F}_\xi(z)$ depends not only on the point z, but also on a realization ξ of the random variable $\xi : \Xi \to \mathbb{R}^d$. The resulting scheme is shown in Algorithm 4.3. Note that we write ξ for both the random variable and its realization. The exact meaning will always be clear from the context.

Assume that the sequences $(w_t)_{t=1}^T = ((u_t, v_t))_{t=1}^T$, $(z_t)_{t=0}^T$, and $(\gamma_t)_{t=1}^T$ as well as the point $\bar{z} = (\bar{x}, \bar{y})$ are generated by Algorithm 4.3. In the course of this method, we use outcomes of independent identically distributed random variables $\xi_t : \Xi \to \mathbb{R}^d$. From now on, we write $\xi_{[t]}$ for the sequence $(\xi_k)_{k=1}^t$, and $(\Xi^t, \mathcal{F}^t, \mathbb{P}^t)$ for the product probability space $(\otimes_{i=1}^t \Xi, \otimes_{i=1}^t \mathcal{F}, \otimes_{i=1}^t \mathbb{P})$.

Algorithm 4.3 Mirror-Prox method with noisy first-order information (adapted from [JNT08])

1: Choose the number of iterations $T \geq 1$.
2: Choose independent identically distributed random variables ξ_t, where $1 \leq t \leq 2T$.
3: Choose $d_{Q_1 \times Q_2}$ as in (4.8).
4: Set $z_0 = (x_0, y_0) = c(d_{Q_1 \times Q_2}) \in Q_1^0 \times Q_2^0$.
5: **for** $1 \leq t \leq T$ **do**
6:     Choose $\gamma_t > 0$ such that

$$\gamma_t \leq \frac{1}{\sqrt{2}L}. \tag{4.13}$$

7:     Observe realizations ξ_{2t-1} and ξ_{2t} of the random variables ξ_{2t-1} and ξ_{2t}, respectively.
8:     Compute $\hat{F}_{\xi_{2t-1}}(z_{t-1})$ and set $w_t = \mathrm{Prox}_{Q_1 \times Q_2, z_{t-1}}(\gamma_t \hat{F}_{\xi_{2t-1}}(z_{t-1}))$.
9:     Determine $\hat{F}_{\xi_{2t}}(w_t)$ and set $z_t = \mathrm{Prox}_{Q_1 \times Q_2, z_{t-1}}(\gamma_t \hat{F}_{\xi_{2t}}(w_t))$.
10: **end for**
11: Return $\bar{z} := \left(\sum_{t=1}^{T} \gamma_t\right)^{-1} \sum_{t=1}^{T} \gamma_t w_t$.

Note that w_t depends on the realizations $\xi_{[2t-1]}$. For notational ease, we omit $\xi_{[2t-1]}$ and write w_t instead of the formally correct $w_t(\xi_{[2t-1]})$. The element w_t constitutes a realization of the random variable $w_t : \Xi^{2t-1} \to Q$. Similarly, z_t is an outcome of the random variable $z_t : \Xi^{2t} \to Q$, and, consequently, Algorithm 4.3 returns a realization $\bar{z}$ of the random variable

$$\bar{z} = \frac{\sum_{t=1}^{T} \gamma_t w_t}{\sum_{t=1}^{T} \gamma_t} : \Xi^{2T-1} \to Q.$$

Note that Algorithm 4.3 corresponds exactly to the Mirror-Prox algorithm with erroneous information discussed in Subsection 3.1 in [JNT08], except that we are allowed to use slightly larger step-sizes than Juditsky et al. This is due to the fact that we consider a slightly less general setting as they do. More precisely, they assume that F satisfies

$$\left\| F(z) - F(z') \right\|_{\mathbb{R}^n \times \mathbb{R}^m, *} \leq \bar{L} \left\| z - z' \right\|_{\mathbb{R}^n \times \mathbb{R}^m} + \bar{M} \qquad \forall \, z, z' \in Q$$

for some given constants $\bar{L}, \bar{M} \geq 0$. However, in the setting of this thesis, the constant $\bar{M}$ vanishes and $\bar{L} := L$. In particular, the operator F does not need to be associated with Saddle-Point Problem (4.1) in their setting.

Let us now extend Theorem 4.3 to the current setting. This result requires the definition of the following entities:

$$
\begin{aligned}
\mu_{z_{t-1}} &:= \mathbb{E}_{\mathbb{P}}\left[\hat{F}_{\xi_{2t-1}}(z_{t-1})|\ \xi_{[2t-2]}\right] - F(z_{t-1}), \\
\mu_{w_t} &:= \mathbb{E}_{\mathbb{P}}\left[\hat{F}_{\xi_{2t}}(w_t)|\ \xi_{[2t-1]}\right] - F(w_t), \\
\sigma_{z_{t-1}} &:= \hat{F}_{\xi_{2t-1}}(z_{t-1}) - \mathbb{E}_{\mathbb{P}}\left[\hat{F}_{\xi_{2t-1}}(z_{t-1})|\ \xi_{[2t-2]}\right], \\
\sigma_{w_t} &:= \hat{F}_{\xi_{2t}}(w_t) - \mathbb{E}_{\mathbb{P}}\left[\hat{F}_{\xi_{2t}}(w_t)|\ \xi_{[2t-1]}\right],
\end{aligned}
\tag{4.14}
$$

where $1 \leq t \leq T$ and $\mathbb{E}_{\mathbb{P}}\left[\hat{F}_{\xi_1}(z_0)|\ \xi_{[0]}\right] := \mathbb{E}_{\mathbb{P}}\left[\hat{F}_{\xi_1}(z_0)\right]$. Note that $\sigma_{z_{t-1}}$ and σ_{w_t} are martingale differences; see Appendix A for the definition of martingale difference sequences and their main properties. Recall from Definition 2.27 that $\Omega_{Q_1 \times Q_2} := \max_{z,z' \in Q_1 \times Q_2} \|z - z'\|_{\mathbb{R}^n \times \mathbb{R}^m}$.

Theorem 4.4 (Adapted version of Theorem 3.2 in [JNT08]) *Define the random variable*

$$
\begin{aligned}
\epsilon &:= 1 + \sqrt{2}\left\|\sum_{t=1}^{T}\gamma_t \sigma_{w_t}\right\|_{\mathbb{R}^n \times \mathbb{R}^m,*} + \Omega_{Q_1 \times Q_2}\sum_{t=1}^{T}\gamma_t \|\mu_{w_t}\|_{\mathbb{R}^n \times \mathbb{R}^m,*} \\
&\quad + 2\sum_{t=1}^{T}\gamma_t^2\left(\|\sigma_{w_t} - \sigma_{z_{t-1}}\|_{\mathbb{R}^n \times \mathbb{R}^m,*}^2 + \|\mu_{w_t} - \mu_{z_{t-1}}\|_{\mathbb{R}^n \times \mathbb{R}^m,*}^2\right).
\end{aligned}
$$

Then,

$$
\mathbb{E}_{\mathbb{P}^{2T}}\left[\overline{\phi}(\bar{x}) - \underline{\phi}(\bar{y})\right] \leq \frac{1}{\sum_{t=1}^{T}\gamma_t}\mathbb{E}_{\mathbb{P}^{2T}}\left[\epsilon\right].
$$

Proof: Let us set $(w_t)_{t=1}^{T} = ((u_t, v_t))_{t=1}^{T}$ and choose $p = (q, r) \in Q_1 \times Q_2$. Let $1 \leq t \leq T$. By Definitions (4.14), we have the expressions:

$$
F(z_{t-1}) = \hat{F}_{\xi_{2t-1}}(z_{t-1}) - \sigma_{z_{t-1}} - \mu_{z_{t-1}} \text{ and } F(w_t) = \hat{F}_{\xi_{2t}}(w_t) - \sigma_{w_t} - \mu_{w_t}.
$$

Combining these relations with the linearity of F and the definition of $\overline{\phi}$ and $\underline{\phi}$, we can show that:

$$
\begin{aligned}
\sum_{t=1}^{T}\gamma_t\left(\overline{\phi}(\bar{x}) - \underline{\phi}(\bar{y})\right) &\leq \sum_{t=1}^{T}\gamma_t\left(\phi(u_t, r) - \phi(q, v_t)\right) \\
&= \sum_{t=1}^{T}\gamma_t\langle F(w_t), w_t - p\rangle \\
&= \sum_{t=1}^{T}\gamma_t\left\langle\hat{F}_{\xi_{2t}}(w_t) - \sigma_{w_t} - \mu_{w_t}, w_t - p\right\rangle. \tag{4.15}
\end{aligned}
$$

Note that

$$\left\| \hat{F}_{\xi_{2t}}(w_t) - \hat{F}_{\xi_{2t-1}}(z_{t-1}) \right\|_{\mathbb{R}^n \times \mathbb{R}^m, *}^2$$

$$\leq \; 2 \left\| F(w_t) - F(z_{t-1}) \right\|_{\mathbb{R}^n \times \mathbb{R}^m, *}^2$$

$$+ 2 \left\| \sigma_{w_t} - \sigma_{z_{t-1}} + \mu_{w_t} - \mu_{z_{t-1}} \right\|_{\mathbb{R}^n \times \mathbb{R}^m, *}^2$$

$$\leq \; 2L^2 \left\| w_t - z_{t-1} \right\|_{\mathbb{R}^n \times \mathbb{R}^m}^2$$

$$+ 2 \left\| \sigma_{w_t} - \sigma_{z_{t-1}} + \mu_{w_t} - \mu_{z_{t-1}} \right\|_{\mathbb{R}^n \times \mathbb{R}^m, *}^2 ,$$

where the concluding inequality is due to the Lipschitz continuity of F. Applying the same arguments as in the proof of Theorem 4.3, that is, Lemma 4.2, Definition (2.26), and Equation (4.9), as well as exploiting the above inequality, we can verify that:

$$\sum_{t=1}^{T} \gamma_t \left\langle \hat{F}_{\xi_{2t}}(w_t), w_t - p \right\rangle$$

$$\leq \; 1 + \sum_{t=1}^{T} \frac{\gamma_t^2}{2} \left\| \hat{F}_{\xi_{2t}}(w_t) - \hat{F}_{\xi_{2t-1}}(z_{t-1}) \right\|_{\mathbb{R}^n \times \mathbb{R}^m, *}^2$$

$$- \frac{1}{2} \sum_{t=1}^{T} \left\| w_t - z_{t-1} \right\|_{\mathbb{R}^n \times \mathbb{R}^m}^2$$

$$\leq \; 1 + \sum_{t=1}^{T} \gamma_t^2 \left\| \sigma_{w_t} - \sigma_{z_{t-1}} + \mu_{w_t} - \mu_{z_{t-1}} \right\|_{\mathbb{R}^n \times \mathbb{R}^m, *}^2$$

$$+ \sum_{t=1}^{T} \left(\gamma_t^2 L^2 - \frac{1}{2} \right) \left\| w_t - z_{t-1} \right\|_{\mathbb{R}^n \times \mathbb{R}^m}^2$$

$$\leq \; 1 + \sum_{t=1}^{T} \gamma_t^2 \left\| \sigma_{w_t} - \sigma_{z_{t-1}} + \mu_{w_t} - \mu_{z_{t-1}} \right\|_{\mathbb{R}^n \times \mathbb{R}^m, *}^2$$

$$\leq \; 1 + 2 \sum_{t=1}^{T} \gamma_t^2 \left\| \sigma_{w_t} - \sigma_{z_{t-1}} \right\|_{\mathbb{R}^n \times \mathbb{R}^m, *}^2$$

$$+ 2 \sum_{t=1}^{T} \gamma_t^2 \left\| \mu_{w_t} - \mu_{z_{t-1}} \right\|_{\mathbb{R}^n \times \mathbb{R}^m, *}^2 , \tag{4.16}$$

where the second-last inequality is due to (4.13). In addition, we have:

$$\sum_{t=1}^{T} \gamma_t \left\langle \sigma_{w_t} + \mu_{w_t}, p - w_t \right\rangle$$

$$= \; \sum_{t=1}^{T} \gamma_t \left\langle \sigma_{w_t}, p - c(d_{Q_1 \times Q_2}) \right\rangle$$

$$+ \sum_{t=1}^{T} \gamma_t \left(\left\langle \sigma_{w_t}, c(d_{Q_1 \times Q_2}) - w_t \right\rangle + \left\langle \mu_{w_t}, p - w_t \right\rangle \right)$$

$$
\begin{aligned}
\leq \quad & \Omega_V(d_{Q_1 \times Q_2}) \left\| \sum_{t=1}^{T} \gamma_t \sigma_{w_t} \right\|_{\mathbb{R}^n \times \mathbb{R}^m, *} \\
& + \sum_{t=1}^{T} \gamma_t \left(\langle \sigma_{w_t}, c(d_{Q_1 \times Q_2}) - w_t \rangle + \Omega_{Q_1 \times Q_2} \| \mu_{w_t} \|_{\mathbb{R}^n \times \mathbb{R}^m, *} \right) \\
= \quad & \sqrt{2} \left\| \sum_{t=1}^{T} \gamma_t \sigma_{w_t} \right\|_{\mathbb{R}^n \times \mathbb{R}^m, *} \\
& + \sum_{t=1}^{T} \gamma_t \left(\langle \sigma_{w_t}, c(d_{Q_1 \times Q_2}) - w_t \rangle + \Omega_{Q_1 \times Q_2} \| \mu_{w_t} \|_{\mathbb{R}^n \times \mathbb{R}^m, *} \right).
\end{aligned}
$$

It remains to combine the above inequality with (4.15) and (4.16), to take expectations on both sides of the resulting inequality, and to observe that $\mathbb{E}_{\mathbb{P}2T} \left[\langle \sigma_{w_t}, c(d_{Q_1 \times Q_2}) - w_t \rangle \right]$ vanishes as σ_{w_t} is a martingale difference and as w_t depends only on $\xi_{[2t-1]}$. ∎

4.2 Interior-Point methods

"The introduction of polynomial-time Interior-Point methods is one of the most remarkable events in the development of Mathematical Programming in the 1980's. The first method of this family was suggested for Linear Programming by the landmark paper of Karmarkar. An excellent complexity result [...] made this work a sensation and subsequently inspired very intensive and fruitful studies."

(Yurii Nesterov and Arkadi Nemirovski, 1993; see Page 1 in [NN93])

In this section, we give a brief review of Interior-Point methods. We do not intend to explain them in full detail, but rather to present their underlying conceptual ideas, which will be sufficient to understand the advantages and disadvantages of these algorithms regarding the tractability of (large-scale) structured optimization problems. We refer to the books of Nesterov and Nemirovski [NN93] and Renegar [Ren01] for an extensive discussion of these methods.

Interior-Point schemes are based on the Newton method. The Newton method is a powerful algorithm for solving convex optimization problems with linear equality constraints and with a twice differentiable objective function. Provided that we start the method close enough from an optimal solution, it convergences quadratically to this point; see Section 1.2.4 in [Nes04] and Section 10.2 in [BV04] for a detailed discussion.

Roughly speaking, the Newton method approximates $\min \{ f(x) : Ax = b \}$, where f is twice differentiable and convex, $A \in \mathbb{R}^{m \times n}$, and $b \in \mathbb{R}^m$, by a sequence of convex quadratic optimization problems. Given a feasible point x_t, the corresponding quadratic auxiliary problem is defined by the second-order Taylor approximation of f around x_t:

$$\min_{x:Ax=b} \left\{ f(x_t) + \nabla f(x_t)^T(x - x_t) + \frac{1}{2}(x - x_t)^T \nabla^2 f(x_t)(x - x_t) \right\}, \quad (4.17)$$

where we denote by $\nabla^2 f(x_t)$ the Hessian of f at x_t. We can rewrite this problem as:

$$\min_{\Delta x} \left\{ \nabla f(x_t)^T \Delta x + \frac{1}{2}\Delta x^T \nabla^2 f(x_t)\Delta x : A\Delta x = 0 \right\}.$$

The next iterate $x_{t+1} := x_t + \Delta x^*$ is then given by the Karush-Kuhn-Tucker (KKT) conditions:

$$\begin{bmatrix} \nabla^2 f(x_t) & A^T \\ A & 0 \end{bmatrix} \begin{bmatrix} \Delta x^* \\ \lambda^* \end{bmatrix} = \begin{bmatrix} -\nabla f(x_t) \\ 0 \end{bmatrix}, \quad (4.18)$$

where λ^* is the associated optimal dual variable. The above system of equations admits a solution $(\Delta x^*, \lambda^*)$ only if the matrix $[\nabla^2 f(x_t)\ A^T; A\ 0]$ is invertible, which can be proved to be the case when f is strongly convex and if A is of full row rank.

We consider now the following the optimization problem:

$$\varphi^* = \min_x \left\{ \langle c, x \rangle : x \in Q,\ Ax = b \right\}, \quad (4.19)$$

where $A \in \mathbb{R}^{m \times n}$, $b \in \mathbb{R}^m$, $c \in \mathbb{R}^n$, and $Q \subset \mathbb{R}^n$ is a closed and convex set with a non-empty interior.

Evidently, the Newton method is not directly applicable to Problem (4.19) because of the constraint $x \in Q$. The basic idea of Interior-Point methods is to replace (4.19) by a sequence of problems which can all be solved individually by the Newton method. For this, we include the intractable constraints in the objective function. Ideally, we would like to have a problem reformulation as

$$\varphi^* = \min_x \left\{ \langle c, x \rangle + \chi_Q(x) : Ax = b \right\},$$

where $\chi_Q(x) = 0$ if $x \in Q$ and $\chi_Q(x) = +\infty$ otherwise. However, this objective function is not differentiable any more, meaning that we cannot apply the Newton method to the above optimization problem. Interior-Point methods resolve this issue by replacing χ_Q by a parametrized barrier function F_Q/μ, where $\mu > 0$. We call $F_Q : \mathbb{R}^n \to \mathbb{R} \cup \{+\infty\}$ a **barrier function for the set Q** if it fulfills the following conditions:

1. $\operatorname{dom} F_Q = \operatorname{int}(Q)$;

2. it is smooth and strictly convex on its domain;

3. it goes to infinity when its argument approaches the boundary of Q, that is, if $x := \lim_{i \to \infty} x_i \in Q \setminus \operatorname{int}(Q)$ with $x_i \in \operatorname{int}(Q)$ for any i, then $\lim_{i \to \infty} F_Q(x_i) = +\infty$;

see Section 6.2.2 in [BTN01]. Given the parameter $\mu > 0$ and a barrier function F_Q for the set Q, the corresponding auxiliary problem is now of the form:

$$\varphi^*(\mu) = \min_x \left\{ \langle c, x \rangle + \frac{1}{\mu} F_Q(x) : Ax = b \right\}. \tag{4.20}$$

Under very mild assumptions, there exists a unique optimal solution to this optimization problem, which we denote by $x^*(\mu)$. We solve this problem for increasing values of μ, as $\langle c, x^*(\mu) \rangle$ converges to φ^* when μ goes to infinity. As Theorem 4.2.7 in [Nes04] shows, it holds that $\langle c, x^*(\mu) \rangle - \varphi^* \leq \mathcal{O}(1/\mu)$. In particular, the increase of μ is a direct measure of the progress of the algorithm.

The curve $\mu \mapsto x^*(\mu)$ is called the **central path**. Interior-Point methods follow this path approximately. Every Auxiliary Problem (4.20) is approximately solved by the Newton method. In order to ensure not only that the steps are well-defined, but also that the approximate solution for a certain μ can be used as starting point for the Newton method with a larger μ, we impose some extra conditions on F_Q. These extra conditions, which ensure that we can increase μ by a constant at every step, describe some relations between the first, the second, and the third derivative of F_Q. If F_Q satisfies these extra requirements, we call F_Q a **ν-self-concordant barrier function**, where $\nu > 0$ is some parameter appearing in the aforementioned relations. A complete description of these conditions would be beyond the scope of this thesis. We refer the interested reader to the book of Nesterov [Nes04]. However, we remark that the self-concordance parameter $\nu > 0$ is the dominating factor of the analytical complexity of Interior-Point methods:

Theorem 4.5 (Theorem 4.2.8 in [Nes04]) *Interior-Point methods require at most*

$$\mathcal{O}\left(\sqrt{\nu} \ln \left[\frac{\nu}{\epsilon} \right] \right)$$

Newton steps in order to compute a feasible ϵ-solution to the Initial Problem (4.19).

Because of their importance for this thesis, let us present the following standard ν-self-concordant barrier functions.

Example 4.1 (Positive orthant; see Section 4.3.2 in [Nes04]) *The function*

$$F_{\mathcal{LP}} : \mathbb{R}^n \to \mathbb{R} \cup \{+\infty\} : x \mapsto \begin{cases} -\sum_{i=1}^n \ln(x_i), & \text{if } x \in \mathbb{R}^n_{>0} \\ +\infty, & \text{otherwise} \end{cases}$$

is a ν-self-concordant barrier function for the set $Q = \mathbb{R}^n_{\geq 0}$ with $\nu = n$. When we apply Interior-Point methods to linear optimization problems, we use $F_{\mathcal{LP}}$ as ν-self-concordant barrier function.

Example 4.2 (Section 4.3.3 in [Nes04]) *Recall that we denote by $\mathcal{S}^n$ the space of real symmetric $(n \times n)$-matrices and that we write*

$$\lambda_n(X) \geq \ldots \geq \lambda_1(X)$$

for the eigenvalues of any symmetric matrix $X \in \mathcal{S}^n$. The function

$$F_{\mathcal{SDP}} : \mathcal{S}^n \to \mathbb{R} \cup \{+\infty\} : X \mapsto \begin{cases} -\sum_i^n \ln(\lambda_i(X)), & \text{if } \lambda_1(X) > 0 \\ +\infty, & \text{otherwise} \end{cases}$$

is an n-self-concordant barrier function for $Q = \{X \in \mathcal{S}^n : \lambda_1(X) \geq 0\}$. We will use this function in Chapter 6, when we apply Interior-Point methods to semidefinite optimization problems.

Clearly, the following question remains: what is the cost of a Newton step? That is, what is the iteration cost? The reader may anticipate that a Newton step can become critically costly for large-scale practical applications, as we need to resolve the System (4.18) of linear equations at every iteration. We discuss this issue in full detail in Section 6.2, where we apply Interior-Point methods to semidefinite optimization problems.

Part II

A new perspective on the Hedge algorithm

Chapter 5

Hedge algorithm and Dual Averaging schemes

IN this chapter, we highlight the broad applicability of Dual Averaging methods by embedding the Hedge algorithm in the framework of these methods. The Hedge algorithm was introduced by Freund and Schapire [FS97] and encompasses many well-known schemes in Machine Learning and in Data Mining. As Freund and Schapire showed, this method is for instance related to the now widely used AdaBoost algorithm[1], which was introduced in the same paper [FS97] as the Hedge algorithm. In addition, Multiplicative Weights Update methods are a variation of the Hedge algorithm; see [AHK05] for a survey. Finally, there exists also a matrix version of the Hedge algorithm, which is called the Matrix Multiplicative Weights Update method [AK07]. We discuss this method and its usefulness for solving semidefinite optimization problems in Chapter 7.

The Hedge algorithm can be best understood in the context of the following online allocation problem. We want to invest an amount of money at the stock market in a portfolio consisting of different assets. After each time step, we can modify the current composition of the portfolio. The Hedge algorithm defines an update strategy for the portfolio such that the average performance that we achieve is not much worse than the average performance of the most favorable investment product. The portfolio update rule is based on the current loss (or gain) that is associated with every investment product.

---

[1] The AdaBoost algorithm is one of the top ten Data Mining algorithms. These top ten algorithms were identified by the IEEE International Conference on Data Mining in December 2006; see [WKR$^+$07] for more details.

A precise description of the Hedge algorithm is given in Section 5.1.

In Section 5.2, we propose an alternative viewpoint on the Hedge algorithm, namely we show that the Hedge algorithm is a Dual Averaging scheme. It is well-known that the Hedge algorithm can be interpreted as a Mirror-Descent scheme with an entropy-type distance-generating function; see for instance Chapter 11 in [CBL06]. However, this interpretation has two drawbacks. First, Mirror-Descent schemes require the definition of a convex objective function; see Section 3.3. In this setting, the current loss of the investment products corresponds to a subgradient of this objective function. However, modeling the performance of a portfolio with a static objective function, even when we allow random losses, is at best questionable. As the last financial crisis has shown, significant sudden changes in the performance of an investment product can appear. Having a static objective function forbids us to deal with these sudden shocks. Second, we observed in Section 3.3 that Mirror-Descent schemes need to consider subgradients with more weight the earlier they appear in order to ensure convergence. However, common sense dictates that recent losses contain more relevant information on the future development of the stock market than losses occurred years ago.

Dual Averaging schemes get rid of the two deficiencies at the same time. When applied to the given allocation problem, Dual Averaging schemes do not make any assumptions on the construction of the losses. For instance, they can be chosen in adversarial way with respect to the current portfolio, they can be randomly generated, or – which reflects some of the latest events at the stock market more accurately – their construction rule may dynamically change. In addition, in Dual Averaging schemes, we can give more weight to the latest losses, which allows us to react much faster to significant changes in the market behavior.

Based on this alternative interpretation of the Hedge algorithm, we give in Section 5.2 three modifications of the Hedge algorithm, namely the Optimal Hedge algorithm, the Optimal Time-Independent Hedge algorithm, and the Optimal Aggressive Hedge algorithm. All these methods have convergence results that are better or at least as good as the convergence guarantee for the vanilla Hedge algorithm. The Optimal Hedge algorithm has the same form as the original Hedge algorithm, except that all method parameters are chosen in an optimal way. The Optimal Time-Independent Hedge algorithm requires less a priori information than the Optimal Hedge algorithm. Finally, the Optimal Aggressive Hedge algorithm considers losses as more relevant the later they appear. Numerical results in Section 5.3 show that all these alternative methods perform better than the vanilla Hedge algorithm in practice. More interestingly, using the Optimal Aggressive Hedge algorithm, we end up with an average benefit that is even better than the profit of the most favorable single investment product, provided that the losses incur shocks reverting the performance of assets. This effect would not have been possible with a static objective function.

Contributions: It is known that the Hedge algorithm is a Mirror-Descent scheme. However, as we pointed out above, this interpretation has two significant drawbacks, which can be resolved by interpreting the Hedge algorithm as an instance of the more general Dual Averaging schemes. To the best of our knowledge, this much more general view on the Hedge algorithm is new. The refinements of the Hedge algorithm presented in Section 5.2, namely the Optimal Hedge algorithm, the Optimal Time-Independent Hedge algorithm, and the Optimal Aggressive Hedge algorithm are original.

Relevant literature: We use the reference [FS97]. The material presented in this chapter is taken from [BB10, BB11].

5.1 The Hedge algorithm

The problem the Hedge algorithm aims at solving can be described as follows. We assume that we want to invest a certain amount of money at the stock market. We have at our disposal a basket of n investment products such as shares, currencies, gold, raw materials, real estates, and so on. Let us denote by $x_{t,i} \geq 0$ the share of the initial amount of money that we invest in product i at time t, where $i = 1, \ldots, n$ and $t \geq 0$. We suppose that we invest the whole amount of money, that is, we assume $x_t \in \Delta_n$ for all $t \geq 0$. At every time step $t \geq 0$, we evaluate the loss (or gain) $\ell_{t,i}$ corresponding to the investment product i, where we assume $\ell_{t,i} \in [-\mu, \rho]$ for every $t \geq 0$ and any $i = 1, \ldots, n$. Thus, given the portfolio x_t at time t, we suffer a loss of $\langle \ell_t, x_t \rangle$ at this time step. The Hedge algorithm defines now a portfolio update strategy such that the averaged loss $\sum_{t=0}^{T-1} \langle \ell_t, x_t \rangle / T$ that we face after $T \in \mathbb{N}$ rounds is not much worse than the averaged total loss $\min_{1 \leq i \leq n} \sum_{t=0}^{T-1} \ell_{t,i} / T$ of the investment product with the best performance.

The Hedge scheme evaluates the losses through a decreasing score function $U : [-\mu, \rho] \to (0, 1]$. We focus on score functions of the form $U(z) = \gamma^{az+b}$, where $\gamma \in (0, 1)$, $a > 0$, and $b \in \mathbb{R}$ are some parameters whose choices we discuss in detail afterwards. The Hedge algorithm assigns a weight $w_{t,i}$ to every investment product $1 \leq i \leq n$ and for every time step $t \geq 0$. The current weight of investment product i depends on its initial weight and on its performance in the past. Concretely, it is defined as $w_{t+1,i} := w_{t,i} U(\ell_{t,i})$. The portfolio x_{t+1} is then given by the normalization of the weight vector w_{t+1}. The full method is shown in Algorithm 5.1.

Given $T \in \mathbb{N}$, let the sequences $(x_t)_{t=0}^{T}$ and $(\ell_t)_{t=0}^{T-1}$ be generated by Algorithm 5.1. Freund and Schapire studied the convergence behavior of this method. In their paper, they considered the situation where $\mu = 0$ and $\rho = 1$.

Algorithm 5.1 Hedge algorithm [FS97]

1: Choose $T \in \mathbb{N}$.
2: Let $\gamma \in (0, 1)$, $a > 0$, $b \in \mathbb{R}$, and $w_0 = (1/n, \ldots, 1/n) \in \mathbb{R}^n$.
3: **for** $t = 0, \ldots, T - 1$ **do**
4:     Set $x_t = w_t / \sum_{i=1}^{n} w_{t,i}$.
5:     Observe loss vector ℓ_t.
6:     Set $w_{t+1,i} = \gamma^{a\ell_{t,i}+b}$ for any $i = 1, \ldots, n$.
7: **end for**

The immediate extension of their reasoning to the more general setting of Algorithm 5.1 yields to the following result.

Theorem 5.1 (Theorem 2 in [FS97]) *With parameters* $a = 1/(\mu + \rho)$ *and* $b = \mu/(\mu + \rho)$, *we have:*

$$\sum_{t=0}^{T-1} (\mu + \langle \ell_t, x_t \rangle) \leq \frac{\mu + \rho}{1 - \gamma} - \frac{\ln(\gamma)}{1 - \gamma} \min_{1 \leq i \leq n} \left(\sum_{t=0}^{T-1} (\mu + \ell_{t,i}) \right).$$

As mentioned in [FS97], the above theorem can be extended to any decreasing score function $U : [-\mu, \rho] \to \mathbb{R}$ that complies with the condition

$$\gamma^{\frac{z+\mu}{\mu+\rho}} \leq U(z) \leq 1 - (1 - \gamma)\frac{z + \mu}{\mu + \rho} \qquad \forall\, z \in [-\mu, \rho]. \tag{5.1}$$

In accordance to Section 2.2 in [FS97], we set

$$\gamma := \frac{1}{\sqrt{2\ln(n)/T} + 1}.$$

We end up with the score function

$$U : [-\mu, \rho] \to \mathbb{R} : z \mapsto \left(\sqrt{2\ln(n)/T} + 1 \right)^{-\frac{z+\mu}{\mu+\rho}}, \tag{5.2}$$

for which one can prove the following statement using Theorem 5.1; see [FS97] for more details on the derivation of this score function.

Corollary 5.1 (Consequence of Lemma 4 in [FS97]) *With the above score function, we have:*

$$\frac{1}{T} \left(\sum_{t=0}^{T-1} \langle \ell_t, x_t \rangle - \min_{1 \leq i \leq n} \sum_{t=0}^{T-1} \ell_{t,i} \right) \leq (\mu + \rho) \left(\frac{\ln(n)}{T} + \sqrt{\frac{2\ln(n)}{T}} \right). \tag{5.3}$$

5.2 The Hedge algorithm is a Dual Averaging method

In this section, we show that we can recast the Hedge algorithm in the framework of Dual Averaging schemes. Based on this alternative view on the Hedge algorithm, we derive three advanced versions of the original method.

Referring to the notation of Algorithm 3.1, we define Q as the $(n-1)$-dimensional standard simplex Δ_n, so that Q encompasses all possible portfolios. We equip $\mathbb{R}^n$ with the norm $\|x\|_1 := \sum_{i=1}^{n} |x_i|$, for which the corresponding dual norm is of the form $\|s\|_\infty := \max_{1 \le i \le n} |s_i|$. In addition, we endow Algorithm 3.1 with the distance-generating function

$$d_{\Delta_n} : \Delta_n \to \mathbb{R} : x \mapsto \ln(n) + \sum_{i=1}^{n} x_i \ln(x_i).$$

Dual Averaging schemes require a parametrized mirror-operator defined as in (3.1). For the given setup, this parametrized mirror-operator is of the form:

$$\pi_{\Delta_n, \beta}(s) = \left(\frac{\exp(s_i/\beta)}{\sum_{j=1}^{n} \exp(s_j/\beta)} \right)_{i=1}^{n} \qquad \forall\, s \in \mathbb{R}^n,\ \forall\, \beta > 0;$$

see Example 2.14. Given a loss vector $\ell_t \in \mathbb{R}^n$, that is, $\ell_{t,i}$ corresponds to the loss of investment product i at time t, we evaluate this vector component-by-component through an affine function $z \mapsto az + b$, where $a > 0$ and $b \in \mathbb{R}$. We interpret the resulting vector $g_t := (a\ell_{t,i} + b)_{i=1}^{n}$ as the output $\mathcal{G}(x_t)$ of the oracle $\mathcal{G}$ in Algorithm 3.1 for input vector $x_t \in \mathbb{R}^n$. Note that we do not specify any construction rule for the loss vector ℓ_t. For instance, it could be chosen randomly or in an adversarial way with respect to the portfolio x_t. Algorithm 3.1 takes the form shown in Algorithm 5.2 for the given setting, where we express the parametrized mirror-operator $\pi_{\Delta_n, \beta}$ in a form that makes the comparison of the resulting method with the Hedge algorithm rather transparent.

Let us now discuss several strategies for choosing the weights γ_t, the projection parameters β_t, and the affine function $z \mapsto az + b$ in Algorithm 5.2. However, first we observe that the norm of each oracle return $a\ell_t + b$ and the distance-generating function d_{Δ_n} are bounded from above by the quantities

$$M(a, b) := \max\left\{ \left| -a\mu + b \right|, \left| a\rho + b \right| \right\} \qquad \text{and by} \qquad \Omega_{d_{\Delta_n}} := \ln(n),$$

respectively. Let the sequences $(x_t)_{t=0}^{T-1}$ and $(\ell_t)_{t=0}^{T-1}$ be generated by Algorithm 5.2.

Fix $\gamma \in (0, 1)$. If $\beta_t = 1$ for any $0 \le t \le T$ and $\gamma_t = \ln(1/\gamma)$ for any $0 \le t \le T - 1$, we recover the Hedge algorithm, which shows that this method is a Dual Averaging scheme. In order to avoid any confusion, we refer henceforth to this method as **Original Hedge algorithm**.

Algorithm 5.2 Extended Hedge algorithm

1: Choose $T \in \mathbb{N}$.
2: Choose positive weights $(\gamma_t)_{t=0}^{T-1}$ and a non-decreasing sequence $(\beta_t)_{t=0}^{T}$ of positive projection parameters.
3: Let $a > 0$, $b \in \mathbb{R}$, and $w_0 = (1/n, \ldots, 1/n) \in \mathbb{R}^n$.
4: **for** $t = 0, \ldots, T - 1$ **do**
5:     Set $x_t = w_t / \sum_{i=1}^{n} w_{t,i}$.
6:     Observe loss vector ℓ_t.
7:     Set $w_{t+1,i} = \exp\left(-\frac{\gamma_t(a\ell_{t,i}+b)}{\beta_{t+1}}\right) w_t^{\frac{\beta_{t+1}}{\beta_t}}$ for any $i = 1, \ldots, n$.
8: **end for**

5.2.1 Optimal Hedge algorithm

Given the weights and projection parameters of the Original Hedge algorithm, Theorem 3.1 implies the inequality:

$$\frac{1}{T} \max_{x \in \Delta_n} \sum_{t=0}^{T-1} \langle \ell_t, x_t - x \rangle$$
$$\leq \frac{1}{aT \ln(1/\gamma)} \left(\ln(n) + \frac{T \ln^2(1/\gamma) \max\left\{ (-a\mu + b)^2, (a\rho + b)^2 \right\}}{2} \right).$$

This bound is minimized by parameters γ^*, a^*, and b^* that satisfy

$$\gamma^* = \exp\left(-\frac{2}{a^*(\mu + \rho)} \sqrt{\frac{2 \ln(n)}{T}} \right) \qquad \text{and} \qquad b^* = \frac{\mu - \rho}{2} a^*,$$

where $a^* > 0$. We refer to Algorithm 5.2 with the just specified setting as **Optimal Hedge algorithm**, for which the following result holds due to the above inequality.

Corollary 5.2 *If the sequences* $(x_t)_{t=0}^{T-1}$ *and* $(\ell_t)_{t=0}^{T-1}$ *are generated by the Optimal Hedge algorithm, we have:*

$$\frac{1}{T} \max_{x \in \Delta_n} \sum_{t=0}^{T-1} \langle \ell_t, x_t - x \rangle \leq (\mu + \rho) \sqrt{\frac{\ln(n)}{2T}}.$$

This result improves Bound (5.3) by the additive quantity $(\mu + \rho) \ln(n)/T$ and by a multiplicative factor of 2. Note that the resulting score function $U(z) = (\gamma^*)^{a^* z + b^*}$ does not comply with Condition (5.1). Therefore, neither Theorem 5.1 nor its extension can be used to establish the above bound.

5.2.2 Optimal Time-Independent Hedge algorithm

The update parameter γ depends on the number of iterations T in both algorithms, the Original Hedge algorithm with the Score Function (5.2) suggested by Freund and Schapire [FS97] and the Optimal Hedge algorithm. However, when investing money at the stock market, we might not want to fix the number of times that we adapt the portfolio in advance. We thus need an update parameter that is independent of T. Adapting Nesterov's Strategy (3.4), we choose $\gamma \in (0,1)$ and set $\gamma_t := \ln(1/\gamma)$, $\beta_0 = 1$, and $\beta_{t+1} = \sum_{k=0}^{t-1} 1/\beta_k$ for any $t \geq 0$. Applying Theorem 3.1, we obtain for any $T \geq 1$:

$$
\frac{1}{T} \max_{x \in \Delta_n} \sum_{t=0}^{T-1} \langle \ell_t, x_t - x \rangle
$$
$$
\leq \frac{\beta_T}{aT \ln(1/\gamma)} \left(\ln(n) + \frac{\ln^2(1/\gamma) \max \left\{ (-a\mu + b)^2 , (a\rho + b)^2 \right\}}{2} \right).
$$

We minimize the right-hand side of the above inequality, that is, we choose $a^* > 0$ and set

$$
\gamma^* = \exp\left(-\frac{2\sqrt{2\ln(n)}}{a^*(\mu + \rho)} \right) \qquad \text{and} \qquad b^* = \frac{\mu - \rho}{2} a^*.
$$

Exploiting Lemma 3.1, we obtain for the resulting method, which we refer to as the **Optimal Time-Independent Hedge algorithm**, the following upper bound on the regret.

Corollary 5.3 *Assume that the sequences* $(x_t)_{t=0}^{T-1}$ *and* $(\ell_t)_{t=0}^{T-1}$ *are generated by the Optimal Time-Independent Hedge algorithm. Then,*

$$
\frac{1}{T} \max_{x \in \Delta_n} \sum_{t=0}^{T-1} \langle \ell_t, x_t - x \rangle \leq (\mu + \rho) \left(\frac{1}{(1+\sqrt{3})\,T} + \sqrt{\frac{2}{T}} \right) \sqrt{\frac{\ln(n)}{2}}.
$$

We observe that the right-hand side of the above inequality can be bounded from above by:

$$
(\mu + \rho) \left(\frac{1}{(1+\sqrt{3})\,T} + \sqrt{\frac{2}{T}} \right) \sqrt{\frac{\ln(n)}{2}} \leq 2(\mu + \rho)\sqrt{\frac{\ln(n)}{T}},
$$

which shows that the convergence rates of the Optimal Time-Independent Hedge algorithm and the Optimal Hedge algorithm are of the same order $\mathcal{O}\left([\mu + \rho]\sqrt{\ln(n)/T} \right)$.

5.2.3 Optimal Aggressive Hedge algorithm

The later a loss appears, the more likely it is that this loss vector contains relevant information on the future development of the investment products' performances. We introduce now an alternative version of the Hedge algorithm where we continuously increase the weights of the loss vectors when time proceeds.

For fixed $\gamma \in (0,1)$, we set $\gamma_t = \ln(1/\gamma)(t+1)^2$ and $\beta_t = t^{2.5}$ for any $t \geq 0$. Let $T > 6$. Using the relations

$$\sum_{t=0}^{T-1}(t+1)^2 = \frac{T(T+1)(2T+1)}{6} > \frac{T^3}{3}, \qquad \sum_{t=0}^{T-1}(t+1)^4 \leq \frac{2T^5}{7},$$

and Theorem 3.1, we obtain for Algorithm 5.2 with the just defined parameters:

$$\frac{6}{T(T+1)(2T+1)} \max_{x \in \Delta_n} \sum_{t=0}^{T-1}(t+1)^2 \langle \ell_t, x_t - x \rangle$$

$$< \frac{3}{aT^3} \cdot \frac{T^{2.5}\ln(n)}{\ln(1/\gamma)}$$

$$+ \frac{3}{aT^3} \cdot \frac{1}{2} \sum_{t=0}^{T-1} \frac{(t+1)^4 \ln(1/\gamma) \max\left\{(-a\mu+b)^2, (a\rho+b)^2\right\}}{T^{2.5}}$$

$$\leq \frac{3}{a\sqrt{T}} \left(\frac{\ln(n)}{\ln(1/\gamma)} + \frac{\ln(1/\gamma) \max\left\{(-a\mu+b)^2, (a\rho+b)^2\right\}}{7} \right).$$

The latter quantity is minimized for a^*, b^*, and γ^* satisfying

$$\gamma^* = \exp\left(-\frac{2\sqrt{7\ln(n)}}{a^*(\mu+\rho)}\right) \qquad \text{and} \qquad b^* = \frac{\mu-\rho}{2}a^*$$

with $a^* > 0$. We refer to the resulting method as the **Optimal Aggressive Hedge algorithm**, for which we can rewrite the above inequality in the following form.

Corollary 5.4 *Let $(x_t)_{t=0}^{T-1}$ and $(\ell_t)_{t=0}^{T-1}$ be generated by the Optimal Aggressive Hedge algorithm. Then,*

$$\frac{6}{T(T+1)(2T+1)} \max_{x \in \Delta_n} \sum_{t=0}^{T-1}(t+1)^2 \langle \ell_t, x_t - x \rangle < 3(\mu+\rho)\sqrt{\frac{\ln(n)}{7T}}.$$

While all the previous methods ensure some bounds on the regret, the update strategy defined by the Optimal Aggressive Hedge algorithm yields to an upper on the weighted regret. The weighted regret reflects the time-varying choice of the weights γ_t.

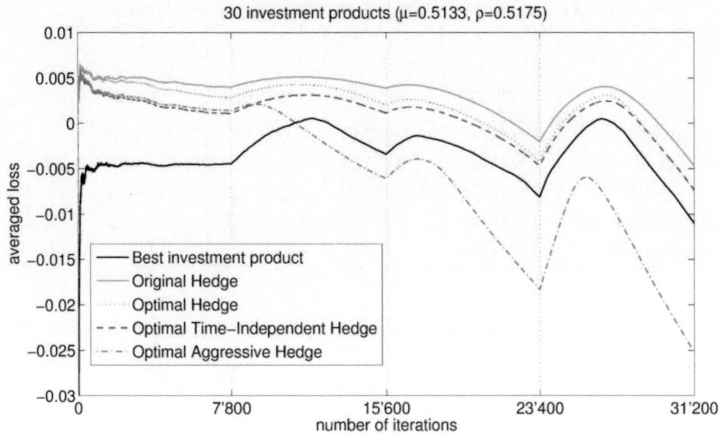

Figure 5.1: *Averaged losses $\sum_{k=1}^{t} \langle \ell_{k-1}, x_{k-1} \rangle / t$, $t = 1, \ldots, T$, achieved by the best investment product (black line), by the Original Hedge algorithm (magenta line), by the Optimal Hedge algorithm (dotted red line), by the Optimal Time-Independent Hedge algorithm (dashed blue line), and by the Optimal Aggressive Hedge algorithm (dashed-dotted green line).*

5.3 Numerical results

We select a pool of $n = 30$ investment products and consider $T = 31'200$ iterations of the methods that we presented. The number T is chosen in such a way that it corresponds to the number of transactions at a stock exchange during four months (twenty trading days of six hours and thirty minutes for one month), provided that there is transaction every minute. The losses $\ell_t \in \mathbb{R}^n$, $t = 0, \ldots, T-1$, are randomly generated. The first $7'800$ losses $(\ell_t)_{t=0}^{7'799}$, that is, the losses observed during the first month, are realizations of a multivariate normally distributed random vector with mean $\bar{\mu}_1$ and covariance matrix Σ. The data $(\bar{\mu}_1, \Sigma)$ is taken from [dat09]. The losses $\left(\ell_{7'800(j-1)+k}\right)_{k=0}^{7'799}$ observed in month j, where $j = 2, 3, 4$, are realizations of a multivariate normally distributed random vector with the same covariance matrix Σ, but with a different mean $\bar{\mu}_j$. In the experiments that are presented, we modify each component $\bar{\mu}_{j-1,i}$ of $\bar{\mu}_{j-1}$ as $\bar{\mu}_{j,i} = a_{j,i}\bar{\mu}_{j-1,i} + b_j$, with b_j small. The coefficient $a_{j,i}$ is negative with an increasing probability as j becomes larger (namely $1/2$, $3/4$, and 1), reverting the performance of

30 investment products ($\mu = 0.5133$, $\rho = 0.5175$)

# iterations	7'800	15'600	23'400	31'200	w.r.t. best product
Best product	−0.0045	−0.0034	−0.0081	−0.0110	−
Original Hedge	0.0040	0.0039	−0.0020	−0.0047	42.7%
Opt. Hedge	0.0028	0.0020	−0.0042	−0.0075	68.2%
Opt. Time-I. H.	0.0010	0.0011	−0.0047	−0.0073	66.4%
Opt. Agg. H.	0.0014	−0.0061	−0.0183	−0.0252	229.1%

Table 5.1: *Averaged losses achieved by the best investment product, by the Original Hedge algorithm, by the Optimal Hedge algorithm, by the Optimal Time-Independent Hedge algorithm, and by the Optimal Aggressive Hedge algorithm after one, two, three, and four months of trading. In the last column, we express the final averaged loss in percentage of the final averaged loss achieved by the best investment product.*

more and more products. The level of perturbation $|a_{j,i}|$ is also increasing as j becomes larger. The experiments are run ten times, and the obtained losses are averaged afterwards.

In Figure 5.1, we show the averaged losses, that is, $\sum_{k=1}^{t} \langle \ell_{k-1}, x_{k-1} \rangle / t$ for any $1 \leq t \leq T$, achieved by the most successful investment product at instant t (obviously, this winning product might change over time), by the Original Hedge algorithm (with Freund and Schapire's score function as described in (5.2)), by the Optimal Hedge algorithm, by the Optimal Time-Independent Hedge algorithm, and by the Optimal Aggressive Hedge algorithm. Note that we show for the Optimal Aggressive Hedge algorithm also the quantity $\sum_{k=1}^{t} \langle \ell_{k-1}, x_{k-1} \rangle / t$, although we use a different weighting in the algorithm and in its theoretical analysis; compare with the last section. In Table 5.1, we give the averaged losses after each month.

We observe that all the extensions of the Hedge algorithm that we suggest in this work significantly outperform its original counterpart. Even more interestingly, the Optimal Aggressive Hedge algorithm achieves an averaged loss that is more than twice better than the averaged loss of the best investment product after four months. The Optimal Aggressive Hedge algorithm outperforms the most successful investment product, as the investment product with the best performance has accumulated a significant loss in an early month. This happens as we switch signs when we perturb the means of the distribution that we use to generate random losses.

Algorithm	Averaged CPU time [sec]
Original Hedge	3.29"
Optimal Hedge	3.12"
Optimal Time-Independent Hedge	3.35"
Optimal Aggressive Hedge	4.05"

Table 5.2: *Averaged CPU time [in seconds] required by the Original Hedge algorithm, by the Optimal Hedge algorithm, by the Optimal Time-Independent Hedge algorithm, and by the Optimal Aggressive Hedge algorithm for performing 31'200 iterations.*

Compared to the every other version of the Hedge algorithm that we suggested in this chapter, the Optimal Hedge algorithm reacts faster and thus more successful to the perturbations. This is due to the increasing weights γ_t, which make losses the more relevant the later they appear. Recall that all the other methods consider the losses as equally important.

As shown in Table 5.2, there is no significant difference with respect to the CPU time required by the different methods for performing 31'200 iterations. All numerical results are run on a computer with two AMD Athlon processors with a CPU of 2.9 GHz and with 3.7 GB of RAM.

Part **III**

Approximately solving large-scale semidefinite optimization problems

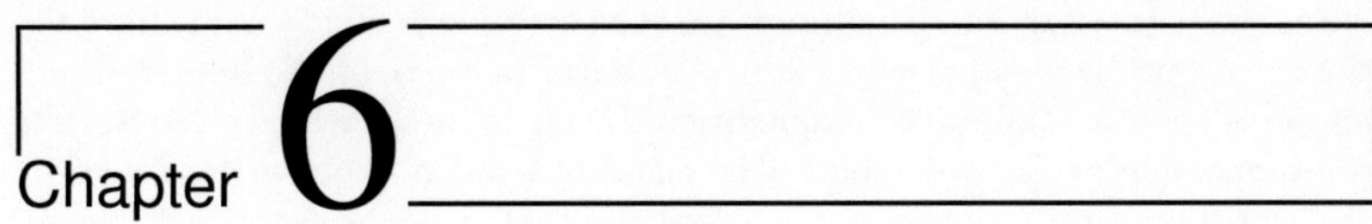

Chapter 6

An introduction to
large-scale Semidefinite Optimization

"Semidefinite Programming is the most exciting development in Mathematical Programming in the 1990's."

(Robert M. Freund, 1999; see Page 1 in [Fre99])

SEMIDEFINITE Optimization, or Semidefinite Programming [SDP], is generally recognized as one of the most fertile developments in Convex Optimization (see for instance [Fre99]), with applications in virtually all fields of engineering. Nowadays, a vast range of real-life optimization problems can be represented or approximated by semidefinite optimization problems, in such different fields as Control [BGFB94], Structural Design [BTN97], or Statistics [BV04], only to name a few. Of particular interest are semidefinite relaxations of hard combinatorial problems; see for instance [GW95, NRT99]. Semidefinite Programming can be seen as a generalization of Linear Programming. As for Linear Programming, the goal in Semidefinite Optimization is to maximize / minimize a linear function over a set of feasible points, which are described by a finite number of inequalities. However, in contrast to Linear Programming, the decision variables are not vectors any more, but symmetric matrices. In Linear Programming, these decision vectors are required to be non-negative. In Semidefinite Programming, we replace this condition by the requirement that all feasible decision matrices are positive-semidefinite, that is, all eigenvalues of every feasible symmetric matrix are supposed to be non-negative.

We give a formal description of semidefinite optimization problems in Section 6.1. Hereby, we restrict ourselves to a subfamily of slightly structured semidefinite optimization problems. This slight structural assumptions will be required in the subsequent chapters, where we apply methods from Part I to these problems.

Standardly, Interior-Point methods are used to solve generic semidefinite optimization problems approximately. As shown in Section 4.2, Interior-Point methods have an analytical complexity of $\mathcal{O}\left(\sqrt{\nu}\ln\left[\nu/\epsilon\right]\right)$, where ϵ denotes the solution accuracy and ν is defined by the ν-self-concordant barrier function. For the semidefinite optimization problems that we consider in this thesis, the parameter ν is given by the sum $m+n$, where m and n correspond to the number of constraints and the size of the decision matrix, respectively. Every iteration of Interior-Point methods consists in carrying out a Newton step. In Section 6.2, we study the cost of such a Newton step, which finally allows us to give a bound on the arithmetic complexity of Interior-Point methods in Semidefinite Optimization.

We conclude this chapter by making two observations regarding the arithmetic complexity of Interior-Point methods in Semidefinite Optimization in Section 6.2. First, Interior-Point methods are of polynomial time. And second, the arithmetic complexity grows – although polynomially – prohibitively fast in the size of the feasible matrices and in the number of constraints for all practical applications that are of very large scale.

Contributions and relevant literature: This chapter is a review of existing definitions and results. It is based on the references [Nes04, NN93].

6.1 Some structured semidefinite optimization problems

We start by presenting a family of slightly structured semidefinite optimization problems. This family includes many important problems such as the semidefinite relaxation of the MaxCut problem; see [GW95] for more details on this relaxation.

Some of the subsequent definitions were introduced (in an explicit or in an implicit form) at an earlier stage of this thesis. However, for the sake of completeness, we summarize them here. Let $\mathcal{M}^n$ be the space of real $(n \times n)$-matrices. The function $\mathrm{Tr} : \mathcal{M}^n \to \mathbb{R} : X \mapsto \sum_{i=1}^{n} X_{ii}$ represents the **trace** of its argument. The **Frobenius scalar product** is denoted by $\langle \cdot, \cdot \rangle_F$, that is,

$$\langle X, Y \rangle_F = \sum_{i,j=1}^{n} X_{ij} Y_{ij} = \mathrm{Tr}(X^T Y) \qquad \forall\, X, Y \in \mathcal{M}^n.$$

We recall that we write $\mathcal{S}^n$ for the $n(n+1)/2$-dimensional vector subspace of **real symmetric matrices** in $\mathcal{M}^n$.

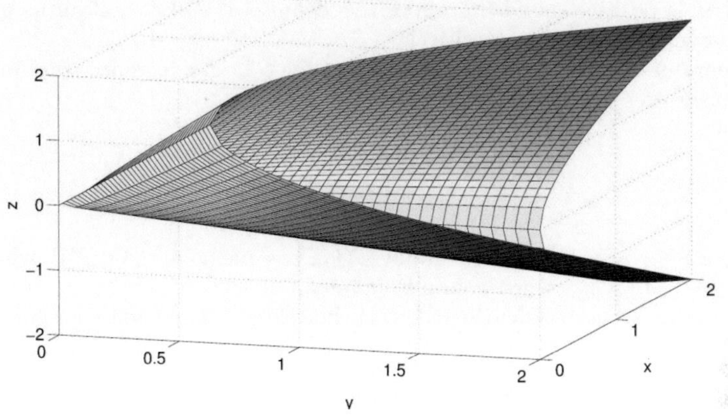

Figure 6.1: *The boundary of $\mathcal{S}_+^2 := \left\{[x\ z; z\ y] : x \geq 0, y \geq 0, |z| \leq \sqrt{xy}\right\}$.*

Definition 6.1 (Positive semidefinite matrix) *Let $X \in \mathcal{S}^n$. We write $X \succeq 0 \ [X \succ 0]$ if X is positive semidefinite [positive definite], that is, if $v^T X v \geq 0$ for every $v \in \mathbb{R}^n$ [resp. $v^T X v > 0$ for each nonzero $v \in \mathbb{R}^n$].*

The set $\mathcal{S}_+^n$ of all positive semidefinite $(n \times n)$-matrices defines a **proper cone**, that is, $\mathcal{S}_+^n \subset \mathbb{R}^{n(n+1)/2}$ is a cone that is closed, convex, solid (A cone is called solid if it has a non-empty interior. The interior of $\mathcal{S}_+^n$ encompasses all positive definite $(n \times n)$-matrices and is thus non-empty.), and pointed (A cone is pointed if it does not contain any line.); see for instance [BV04] for more details. In Figure 6.1, we give a graphical representation of the boundary of $\mathcal{S}_+^2$.

As indicated in the introduction of this chapter, we can alternatively describe the set $\mathcal{S}_+^n$ by the eigenvalues of the considered matrices. Recall that we denote by $\lambda(X)$ the vector formed by the eigenvalues of the matrix $X \in \mathcal{S}^n$. Every $X \in \mathcal{S}^n$ admits an eigendecomposition

$$X = Q(X)\mathrm{Diag}(\lambda(X))Q(X)^T,$$

where $Q(X)$ is a (not necessarily unique) orthogonal matrix of dimension $(n \times n)$ and $\mathrm{Diag}(\lambda(X))$ represents the diagonal matrix whose diagonal is $\lambda(X)$. We observe that all eigenvalues of any matrix $X \in \mathcal{S}^n$ are real. Conventionally, we assume that the eigenvalues $\lambda_n(X) \geq \ldots \geq \lambda_1(X)$ are ordered decreasingly. We write

$$\lambda_{\max}(X) := \lambda_n(X) \qquad \text{and} \qquad \lambda_{\min}(X) := \lambda_1(X)$$

for the largest and the smallest eigenvalue of X, respectively. A symmetric matrix is positive semidefinite [positive definite] if and only if all its eigenvalues are non-negative [positive].

Eigenvalues can be used to define norms for $\mathcal{S}^n$. For instance, the *induced p-norms* are:

$$\|X\|_{(p)} := \sqrt[p]{\sum_{i=1}^{n} |\lambda_i(X)|^p},$$

where $1 \le p < \infty$ and $X \in \mathcal{S}_n$, and $\|X\|_{(\infty)} := \max_{i=1,\dots,n} |\lambda_i(X)|$, which is the limit of $\|X\|_{(p)}$ when p goes to infinity.

In this thesis, we consider the following wide class of semidefinite optimization problems:

$$\begin{aligned}
\varphi^* = \max_{X} \quad & \langle C, X \rangle_F \\
\text{s.t.} \quad & \langle A_j, X \rangle_F \le 1 \qquad \text{for any } j = 1, \dots, m \\
& X \succeq 0,
\end{aligned} \qquad \text{(SDP)}$$

where $A_1, \dots, A_m, C \in \mathcal{S}^n$. Loosely speaking, we call this problem **large-scale** if it involves large matrices (that is, n is large), or if it has many constraints (that is, m is large). Let us verify that we cover indeed a wide class of semidefinite optimization problems by Formulation (SDP). The most general class of semidefinite optimization problems is given by:

$$\begin{aligned}
\max_{Y} \quad & \langle D, Y \rangle_F \\
\text{s.t.} \quad & \mathcal{A}(Y) = b \\
& Y \succeq 0,
\end{aligned} \qquad (6.1)$$

where $D \in \mathcal{S}^p$, $\mathcal{A} : \mathcal{S}^p \to \mathbb{R}^q$ is a linear operator, and $b \in \mathbb{R}^q$. In (SDP), we restrict ourselves to the following subfamily of (6.1):

$$\begin{aligned}
\max_{X \in \mathcal{S}^n, u \in \mathbb{R}^m} \quad & \langle C, X \rangle_F \\
\text{s.t.} \quad & \langle A_j, X \rangle_F + u_j = 1 \qquad \text{for any } j = 1, \dots, m \\
& \begin{bmatrix} X & 0 \\ 0 & \mathrm{Diag}(u) \end{bmatrix} \succeq 0.
\end{aligned}$$

The dual problem of (SDP) takes the following form:

$$\begin{aligned}
\min_{y} \quad & \sum_{j=1}^{m} y_j \\
\text{s.t.} \quad & \sum_{j=1}^{m} y_j A_j \succeq C \\
& y \ge 0.
\end{aligned} \qquad (6.2)$$

The following two duality results are of fundamental importance.

Theorem 6.1 (Weak Duality; see Section 4.2.2 in [NN93]) *Choose a matrix $X \in \mathcal{S}_+^n$ and a vector $y \in \mathbb{R}_{\geq 0}^m$ such that $\max_{1 \leq j \leq m} \langle A_j, X \rangle_F \leq 1$ and $\sum_{j=1}^m y_j A_j \succeq C$. The corresponding duality gap $\langle C, X \rangle_F - \sum_{j=1}^m y_j$ is non-negative.*

Theorem 6.2 (Strong Duality; see Section 4.2.2 in [NN93])
Assume that both the Primal Problem (SDP) and its Dual (6.2) are strictly feasible, that is, there exist an $\bar{X} \succ 0$ such that $\langle A_j, \bar{X} \rangle_F < 1$ for any $1 \leq j \leq m$ and a $\bar{y} \in \mathbb{R}_{\geq 0}^m$ with $\sum_{j=1}^m \bar{y}_j A_j \succ C$. Then, the following statements hold:

1. *Both the primal and the dual problem are solvable and attain their optimal values.*

2. *A primal-dual pair (X, y) of feasible points to (SDP) and (6.2) is optimal if and only if the corresponding duality gap is zero.*

We can easily construct a point $\bar{X}$ which is strictly feasible for (SDP).

Remark 6.1 *We choose $\alpha > 0$ such that $\mathrm{Tr}(A_j) < \alpha$ for any $1 \leq j \leq m$. Recall that we write $\mathbb{I}_n$ for the $(n \times n)$-identity matrix. The matrix $\bar{X} := \mathbb{I}_n/\alpha$ is positive definite and satisfies $\langle A_j, \bar{X} \rangle_F = \mathrm{Tr}(A_j)/\alpha < 1$ for any $1 \leq j \leq m$.*

In order to ensure that Strong Duality holds, we make the following assumption, which shall hold for the rest of this thesis.

Assumption 6.1 *There exists an $\bar{y} \in \mathbb{R}_{\geq 0}^m$ such that $\sum_{j=1}^m \bar{y}_j A_j \succ C$.*

As an immediate consequence, we can validate the following result.

Lemma 6.1 *Choose a matrix $X \succeq 0$ such that $\langle C, X \rangle_F > 0$. Then, $\max_{1 \leq j \leq m} \langle A_j, X \rangle_F > 0$.*

Proof: Assume that there exists a matrix $X \succeq 0$ satisfying both inequalities $\langle C, X \rangle_F > 0$ and $\max_{1 \leq j \leq m} \langle A_j, X \rangle_F \leq 0$. For any $\lambda > 0$, the matrix λX is feasible for (SDP). In particular, the Problem (SDP) is unbounded, which is a contradiction to the assumption that Strong Duality holds.   ∎

6.2 Interior-Point methods in Semidefinite Optimization

In this section, we discuss the cost of a single iteration of Interior-Point methods when applied to Problem (SDP). We follow Chapter 4 in the book of Nesterov [Nes04].

In a first step, we rewrite Problem (SDP) as

$$-\varphi^* = \min_{X,u} \quad \langle -C, X \rangle_F$$
$$\text{s.t.} \quad \langle A_j, X \rangle_F + u_j = 1 \qquad \text{for any } j = 1, \ldots, m \tag{6.3}$$
$$X \succeq 0, \ \operatorname{Diag}(u) \succeq 0.$$

Given the $(m + n)$-self-concordant barrier function

$$F(X, u) = \begin{cases} -\ln(\det(X)) - \sum_{j=1}^{m} \ln(u_j), & \text{if } X \succ 0 \text{ and } u > 0 \\ \infty, & \text{otherwise,} \end{cases}$$

Interior-Point methods replace the above problem by a sequence of auxiliary problems of the form:

$$-\varphi^*(\mu) = \min_{X,u} \quad \langle -C, X \rangle_F + \tfrac{1}{\mu} F(X, u)$$
$$\text{s.t.} \quad \langle A_j, X \rangle_F + u_j = 1 \qquad \text{for any } j = 1, \ldots, m, \tag{6.4}$$

where $\mu > 0$ is some parameter; compare with Section 4.2. In Interior-Point methods, the optimal solution to Problem (6.4) is approximated by carrying out one Newton step; see Section 4.2. Let us determine now the cost of this Newton step. We use the notation

$$Y := \begin{bmatrix} X & 0 \\ 0 & \operatorname{Diag}(u) \end{bmatrix}, \qquad Y_t := \begin{bmatrix} X_t & 0 \\ 0 & \operatorname{Diag}(u_t) \end{bmatrix},$$
$$B_j := \begin{bmatrix} A_j & 0 \\ 0 & e_j e_j^T \end{bmatrix}, \qquad D := \begin{bmatrix} -C & 0 \\ 0 & 0 \end{bmatrix},$$

where $X, X_t \in \mathcal{S}^n$, $u, u_t \in \mathbb{R}^m$, and $e_j \in \mathbb{R}^m$ denotes the j-th unit vector. Given the iterates $(X_t, u_t) \in \operatorname{int}(\mathcal{S}_+^n \times \mathbb{R}_{\geq 0}^m)$ and the parameter $\mu_t > 0$, the corresponding Quadratic Approximation Problem (4.17) takes the form

$$\min_{Y} \quad \langle \mu_t D - Y_t^{-1}, Y - Y_t \rangle_F + \tfrac{1}{2} \langle Y_t^{-1}(Y - Y_t) Y_t^{-1}, Y_t - Y \rangle_F$$
$$\text{s.t.} \quad \langle B_j, Y \rangle_F = 1 \qquad \text{for any } j = 1, \ldots, m$$
$$Y \succ 0,$$

which can be rewritten as

$$\min_{\Delta Y} \quad \langle \mu_t D - Y_t^{-1}, \Delta Y \rangle_F + \tfrac{1}{2} \langle Y_t^{-1} \Delta Y Y_t^{-1}, \Delta Y \rangle_F$$
$$\text{s.t.} \quad \langle B_j, \Delta Y \rangle_F = 0 \qquad \text{for any } j = 1, \ldots, m$$
$$Y_t + \Delta Y \succ 0;$$

see Subsection 4.3.3 in [Nes04]. We use the necessary and sufficient KKT conditions to determine a solution ΔY^* to the latter problem. Due to these conditions, ΔY^* is defined by the following linear equations:

$$\mu_t D - Y_t^{-1} + Y_t^{-1} \Delta Y^* Y_t^{-1} - \sum_{j=1}^{m} \lambda_j^* B_j \;=\; 0$$

$$\langle \Delta Y^*, B_j \rangle_F \;=\; 0 \qquad \forall\, j = 1, \ldots, m; \quad (6.5)$$

see Equations (4.3.8) in [Nes04]. According to Theorem 4.1.14 in [Nes04] or to Theorem 2.2.2 in [NN93], the point $Y^* = Y_t + \Delta Y^*$ belongs to the interior of $\mathcal{S}_+^n \times \mathbb{R}_{\geq 0}^m$, provided that μ_t is chosen appropriately. The first condition of (6.5) can be stated as

$$\Delta Y^* = Y_t \left(\sum_{j=1}^{m} \lambda_j^* B_j - \mu_t D + Y_t^{-1} \right) Y_t,$$

which allows us to rewrite the second group of conditions in (6.5) as

$$\sum_{k=1}^{m} \lambda_k^* \langle B_j, Y_t B_k Y_t \rangle_F = \langle Y_t \left(\mu_t D - Y_t^{-1} \right) Y_t, B_j \rangle_F \qquad \forall\, j = 1, \ldots, m.$$

This system of linear equations can be written in matrix form as $U\lambda^* = d$, where

$$U_{jk} := \langle B_j, Y_t B_k Y_t \rangle_F \qquad \text{and} \qquad d_j := \langle Y_t \left(\mu_t D - Y_t^{-1} \right) Y_t, B_j \rangle_F$$

for any $j, k = 1, \ldots, m$. As shown in [Nes04], we can solve System (6.5) by performing the following operations:

1. **Computation of the matrices $Y_t B_k Y_t$:** As

$$Y_t B_k Y_t = \begin{bmatrix} X_t A_k X_t & 0 \\ 0 & u_{t,k}^2 e_k e_k^T \end{bmatrix} \qquad \forall\, k = 1, \ldots, m,$$

   we need $\mathcal{O}\left(mn^3\right)$ arithmetic operations to compute the m $(n \times n)$-matrices $Y_t B_k Y_t$. Assume now that all matrices A_k are sparse and that S_k denotes the number of non-zero entries of the matrix A_k. We define $S := \max_{1 \leq k \leq m} S_k$. Then, the computation of $X_t A_k$ needs $\mathcal{O}(nS)$ arithmetic operations. However, the resulting matrix might become dense, implying that the evaluation of all matrices $(Y_t B_k) Y_t$, $k = 1, \ldots, m$, requires $\mathcal{O}\left(mn^3\right)$ arithmetic operations, as well. Consequently, sparsity is of no help in this step.

2. **Computation of the matrix U:** Given the $(n \times n)$-matrices $Y_t B_k Y_t$, the computation of the m^2 elements

$$U_{jk} = \langle B_j, Y_t B_k Y_t \rangle_F = \langle A_j, X_t A_k X_t \rangle_F + u_{t,k}^2 \langle e_j, e_k \rangle, \; 1 \leq j, k \leq m,$$

   of the matrix U requires $\mathcal{O}\left(m^2 n^2\right)$ $[\mathcal{O}\left(m^2 S\right)$ if the matrices A_j are sparse] arithmetic operations. Note that U might be dense also for sparse input matrices A_j.

3. **Computation of the vector d:** In order to compute the vector d, we need to evaluate first the expression

$$\mu_t Y_t D Y_t - Y_t = \begin{bmatrix} -\mu_t X_t C X_t - X_t & 0 \\ 0 & -\mathrm{Diag}(u_t) \end{bmatrix},$$

which requires $\mathcal{O}(n^3 + m)$ arithmetic operations. For any $j = 1, \dots, m$, we derive the value of $\langle \mu_t Y_t D Y_t - Y_t, B_j \rangle_F$ by carrying out $\mathcal{O}(n^2)$ arithmetic operations. In total, we thus need $\mathcal{O}\left(n^3 + mn^2\right)$ $\left[\mathcal{O}\left(n^3 + mS\right)\right.$ for sparse matrices A_j] arithmetic operations to compute d.

4. **Computation of the vector $\boldsymbol{\lambda^*}$:** The computation of $\lambda^* = U^{-1} d$ requires $\mathcal{O}(m^3)$ arithmetic operations.

5. **Computation of the matrix $\boldsymbol{\Delta Y^*}$:** We need $\mathcal{O}\left(mn^2\right)$ $\left[\mathcal{O}\left(mS\right)\right.$ for sparse matrices A_j] arithmetic operations to compute

$$\sum_{j=1}^{m} \lambda_j^* B_j = \begin{bmatrix} \sum_{j=1}^{m} \lambda_j^* A_j & 0 \\ 0 & \mathrm{Diag}\left(\lambda^*\right) \end{bmatrix}.$$

Given this matrix, $\mathcal{O}\left(n^3 + m\right)$ arithmetic operations are required for the computation of

$$Y_t \left(\sum_{j=1}^{m} \lambda_j^* B_j \right) Y_t = \begin{bmatrix} X_t \left(\sum_{j=1}^{m} \lambda_j^* A_j \right) X_t & 0 \\ 0 & \mathrm{Diag}\left((\lambda_j^* u_{t,j}^2)_{j=1}^{m} \right) \end{bmatrix}.$$

It remains to subtract $\mu_t Y_t D Y_t - Y_t$ from this matrix, which requires $\mathcal{O}\left(n^2 + m\right)$ arithmetic operations. Note that we have already computed $\mu_t Y_t D Y_t - Y_t$ above and that sparsity of the matrices A_j does not affect the last two complexity estimates.

In total, we thus need $\mathcal{O}\left(mn^3 + m^2 n^2 + m^3\right)$ $\left[\mathcal{O}\left(mn^3 + m^2 S + m^3\right)\right.$ for sparse matrices A_j] arithmetic operations per Newton step.

Combining the cost per iteration with the analytical complexity described in Theorem 4.5 and recalling that F is a ν-self-concordant barrier function with $\nu = m + n$ (see Example 4.2), we have verified the following result.

Theorem 6.3 (Section 4.3.3 in [Nes04]) *Interior-Point* (IP) *methods require*

$$\mathrm{Compl}_{\mathrm{IP}}(\mathrm{SDP}, \epsilon) = \mathcal{O}\left(\sqrt{m+n}\left[mn^3 + m^2 S + m^3\right]\ln\left[(m+n)/\epsilon\right]\right)$$

arithmetic operations to compute a feasible ϵ-solution to Problem (SDP).

As an immediate consequence, we obtain:

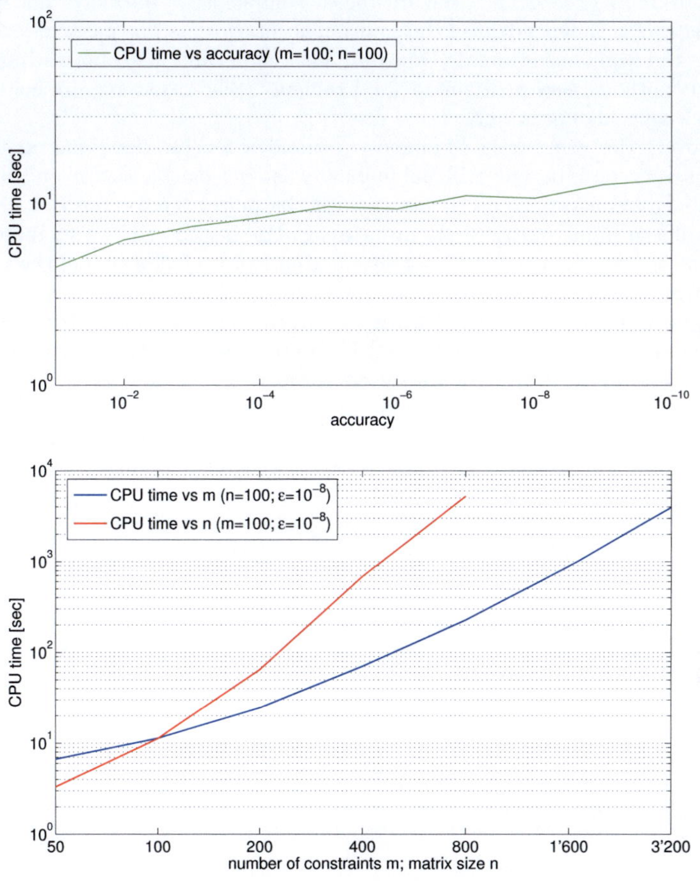

Figure 6.2: *CPU time required by Interior-Point methods (SeDuMi) for solving single random instances of Problem (SDP) with respect to the accuracy ϵ, the number of constraints m, and the matrix size n.*

Corollary 6.2 *Interior-Point methods are of polynomial time on the family of semidefinite optimization problems described in (SDP).*

According to Theorem 6.3, the arithmetic complexity of Interior-Point methods is almost independent of the solution accuracy ϵ, as this parameter influences the worst-case running time only logarithmically. This makes Interior-Point methods very well suited for Problems (SDP) that require solutions with a very high accuracy.

However, the worst-case complexity estimates for Interior-Point methods grows very rapidly, but still polynomially, in the matrix size n and in the number of constraints m: it grows with the power 3.5 for both m and n, even if the input matrices A_j are sparse. For instance, if we increase the matrix size n by one order of magnitude, that is, by a factor of ten, the worst-case running time is multiplied by about $3'100$. We conclude: **in practice, polynomial-time Interior-Point methods are too costly for solving approximately Problems (SDP) if the involved matrices are of very large size or if there are many constraints!**

The above observations are demonstrated by numerical evidence. In Figure 6.2, we plot the CPU time required by Interior-Point methods (SeDuMi[1]) for solving random instances of Problem (SDP) versus the accuracy, the number of constraints, and the matrix size. The numerical experiments are run on a computer with two AMD Athlon processors with a CPU of 2.9 GHz and with 3.7 GB of RAM.

---

[1]SeDuMi is an open-source Matlab toolbox for Semidefinite Optimization; see [Sed, Stu99] for more details.

Chapter 7

A matrix version of the Hedge algorithm in Semidefinite Optimization

In the last chapter, we observed that Interior-Point methods can be used to solve (SDP) in practice if the problem is of medium size, that is, if the problem has a few thousand constraints and if it involves matrices of size a few hundred times a few hundred; see Figure 6.2. In addition, we noticed that Interior-Point methods are a distinguished tool for finding approximate solutions of very high accuracy; SeDuMi, which is one of the most prominent and most frequently used implementations of Interior-Point methods, has a default accuracy of 10^{-8}. However, when it comes to modern practical applications, the needs may be different. On the one hand, practitioners are usually satisfied with solutions with two or three accuracy digits, they typically do not need solutions with an accuracy of 10^{-8}. On the other hand, semidefinite optimization problems arising in practice can be of very large scale. Problematically, the scale of these problems may lie beyond the size of problems that can be successfully handled by Interior-Point methods in practice. We recall here that the running time of Interior-Point methods grows with the order $\mathcal{O}\left(\sqrt{m+n}\left[mn^3 + mS + m^3\right]\ln[m+n]\right)$ with respect to m and n, respectively. These two observations motivate the following conceptual question.

Question 7.1 *We accept to face a moderate complexity increase with respect to the solution accuracy ϵ; say from $\ln[1/\epsilon]$ to $1/\epsilon$ or even to $1/\epsilon^2$. Do algorithms for solving approximately semidefinite optimization problems in the form of (SDP) with lower running time in m and n, say with a worst-case complexity of at most $\mathcal{O}\left((m+n)^3\right)$, exist?*

In Part I of this thesis, we prepared a whole battery of alternative methods (Dual Averaging schemes, Smoothing Techniques, Mirror-Prox methods), which can be exploited to tackle this question. Recall that these algorithms process only dual information (general loss vectors in Dual Averaging schemes, gradients or subgradients in the case of First-Order methods). In particular, these algorithms avoid the costly Newton steps, which makes them an attractive alternative to Interior-Point methods for large-scale problems. However, these alternative methods are not directly applicable to semidefinite optimization problems in the form of (SDP). The Formulation (SDP) does neither comply with Assumption 2.1 on the existence of a closed-form solution to Problem (2.10), nor it has the saddle-point structure which is required by Smoothing Techniques and Mirror-Prox methods. In this and the next chapter, we study different techniques for rewriting Problem (SDP) in a form to which the alternative methods are applicable.

The practical tractability of large-scale semidefinite optimization problems has attracted the interest of many researchers over the last few years. One of the most well-known papers in this research area is due to Arora and Kale. In fact, their work [AK07] was the starting point and motivation of this thesis. They replace Problem (SDP) by a cascade of feasibility problems. Each of these feasibility problems is approximately solved by a Matrix Multiplicative Weights Update method, which can be seen as a matrix version of the Hedge algorithm and can also be embedded in the framework of Dual Averaging schemes. In total, $\tilde{\mathcal{O}}(\ln[1/\epsilon]/\epsilon^2)$ iterations of this Matrix Multiplicative Weights Update method are required, where we use the $\tilde{\mathcal{O}}$-notation to indicate that some other problem parameters such as the scaling of the problem are hidden in this complexity estimate. The cost per iteration is dominated by a matrix exponentiation and by some other computations requiring $\mathcal{O}(mn^2)$ arithmetic operations. Applying standard techniques for the matrix exponentiation, their method needs in total $\tilde{\mathcal{O}}([n^3 + mn^2]\ln[1/\epsilon]/\epsilon^2)$ arithmetic operations to find a feasible ϵ-solution to Problem (SDP), which gives us a first, but rough, answer to the above question. "Rough" in the sense that the quality of their answer heavily depends on the size of the parameters that are hidden in the $\tilde{\mathcal{O}}$-notation. And, as we will establish in this thesis, this complexity result leaves space for improvement with respect to both the solution accuracy ϵ and the matrix size n.

We study the method of Arora and Kale in this chapter. In Section 7.1, we introduce their Matrix Multiplicative Weights Update algorithm and show that this method is a Dual Averaging scheme. In Section 7.2, we first discuss the reformulation of (SDP) as a cascade of feasibility problems. In a second step, we present Arora and Kale's particular instanciation of the Matrix

Multiplicative Weights Update method which can be used to solve these feasibility problems approximately.

Contributions: We point out that the Matrix Multiplicative Weights Update method is a Dual Averaging scheme, which enables us to give an alternative performance analysis of this algorithm.

Relevant literature: We use the paper [AK07] of Arora and Kale.

7.1 The Matrix Multiplicative Weights Update scheme

The Matrix Multiplicative Weights Update algorithm was independently discovered by Arora and Kale [AK07] as well as by Warmuth and Kuzmin [WK06]. It can be interpreted as a generalization of the Hedge algorithm to matrices, where the decision variables are no longer non-negative unitary n-dimensional vectors, but positive semidefinite $(n \times n)$-matrices with trace one.

Let us introduce the Matrix Multiplicative Weights Update method as in the setting presented in [AK07]. We consider the following online matrix version of the usual two-player zero-sum game. At the beginning of round $t \in \mathbb{N}_0$ of the game, the first player chooses a positive semidefinite matrix X_t of trace one. After this action, the second player (who might follow an adversarial strategy), picks a matrix $0 \preceq M_t \preceq \mathbb{I}_n$, where we recall that $\mathbb{I}_n$ denotes the identity matrix in $\mathcal{S}^n$. Any round t of this game is concluded by the following pay-off: the first player pays $\langle M_t, X_t \rangle_F$ to the second player. After $T \in \mathbb{N}$ rounds, the total loss suffered by the first player amounts to $\sum_{t=0}^{T-1} \langle M_t, X_t \rangle_F$. The Matrix Multiplicative Weights Update method defines an update strategy for the matrices $(X_t)_{t \geq 0}$ such that the regret at $T \in \mathbb{N}$, that is,

$$
\begin{aligned}
\mathcal{R}_T &= \sum_{t=0}^{T-1} \langle M_t, X_t \rangle_F - \min_{X \in \Delta_n^M} \left\{ \sum_{t=0}^{T-1} \langle M_t, X \rangle_F \right\} \\
&= \sum_{t=0}^{T-1} \langle M_t, X_t \rangle_F - \lambda_{\min} \left(\sum_{t=0}^{T-1} M_t \right),
\end{aligned}
$$

can be bounded from above by a quantity which is increasing in the order of $\mathcal{O}(\sqrt{T})$. This update strategy is described in Algorithm 7.1.

At every iteration of this method, we are supposed to carry out the computation of $(1-\gamma)^{S_{t+1}}$, where $\gamma \in (0,1)$ and $S_{t+1} \in \mathcal{S}^n$. The matrix $(1-\gamma)^{S_{t+1}}$ is defined as follows. As the matrix S_{t+1} is symmetric, it admits an eigendecomposition $S_{t+1} = Q(S_{t+1}) \mathrm{Diag}(\lambda(S_{t+1})) Q(S_{t+1})^T$. Using this alternative representation of S_{t+1}, we define $(1-\gamma)^{S_{t+1}}$ as

$$
(1-\gamma)^{S_{t+1}} = Q(S_{t+1}) \mathrm{Diag}(\bar{\lambda}) Q(S_{t+1})^T, \qquad \bar{\lambda}_i = (1-\gamma)^{\lambda_i(S_{t+1})} \, \forall \, 1 \leq i \leq n.
$$

Algorithm 7.1 Matrix Multiplicative Weights Update method [AK07]

1: Choose $T \in \mathbb{N}$.
2: Choose $0 < \gamma < 1$, set $X_0 = \mathbb{I}_n/n$, and set $S_0 = 0 \in \mathcal{S}^n$.
3: **for** $0 \leq t \leq T - 1$ **do**
4:     Obtain $0 \preceq M_t \preceq \mathbb{I}_n$ from the second player.
5:     Set $S_{t+1} = S_t + M_t$.
6:     Update $W_{t+1} = (1 - \gamma)^{S_{t+1}}$.
7:     Set $X_{t+1} = W_{t+1}/\mathrm{Tr}(W_{t+1})$.
8: **end for**

In particular, the computation of $(1 - \gamma)^{S_{t+1}}$ requires $\mathcal{O}(n^3)$ arithmetic operations.

In their paper [AK07], Arora and Kale derived an upper bound on $\mathcal{R}_T$ for Algorithm 7.1, provided that the update update parameter γ is strictly smaller than $1/2$. Instead of presenting here their performance study, we interpret Algorithm 7.1 as a Dual Averaging scheme at first and apply Theorem 3.1 afterwards to obtain an upper bound on the regret $\mathcal{R}_T$. Using this alternative view on the Matrix Multiplicative Weights Update method, we can eliminate the extra condition $\gamma < 1/2$, which plays a crucial role in the proof of Arora and Kale.

Consider now Dual Averaging schemes (Algorithm 3.1). We make the following choices in Algorithm 3.1:

1. we set $Q = \Delta_n^M$ and equip $\mathcal{S}^n$ with the induced 1-norm $\|\cdot\|_{(1)}$;

2. we endow Algorithm 3.1 with the distance-generating function

$$d_{\Delta_n^M}(X) = \ln(n) + \sum_{i=1}^{n} \lambda_i(X) \ln(\lambda_i(X)) \qquad \forall\, X \in \Delta_n^M;$$

3. the return of the oracle $\mathcal{G}$ at iteration t corresponds to the matrix M_t chosen by the second player;

4. we use constant step-sizes $\gamma_t = 1$ for any $0 \leq t \leq T - 1$;

5. we set $\beta_t = -1/\ln(1 - \gamma)$ for any $0 \leq t \leq T$.

At every iteration $0 \leq t \leq T - 1$ of Algorithm 3.1, we obtain:

$$X_{t+1} := \pi_{\Delta_n^M,\, -1/\ln(1-\gamma)}\left(-\sum_{\tau=0}^{t} M_\tau\right) = \frac{W_{t+1}}{\mathrm{Tr}(W_{t+1})},$$

where:

$$W_{t+1} \quad := \quad \exp\left(-S_{t+1}/\beta_{t+1}\right) = \exp\left(\ln(1-\gamma)S_{t+1}\right) = (1-\gamma)^{S_{t+1}},$$

$$S_{t+1} \quad := \quad \sum_{\tau=0}^{t} M_{\tau}.$$

We observe that Algorithm 3.1 defines the same sequence $(X_t)_{t=0}^{T}$ of matrices as Algorithm 7.1, that is, the Matrix Multiplicative Weights Update method is a Dual Averaging scheme. By applying Theorem 3.1, we obtain the following result for the Matrix Multiplicative Weights Update method.

Theorem 7.1 *Let the sequences $(X_t)_{t=0}^{T}$ and $(M_t)_{t=0}^{T-1}$ be generated by Algorithm 7.1. Then,*

$$\sum_{t=0}^{T-1} \langle M_t, X_t \rangle_F - \lambda_{\min}\left(\sum_{t=0}^{T-1} M_t\right) \leq -\frac{\ln(n)}{\ln(1-\gamma)} - \frac{\ln(1-\gamma)}{2}\sum_{t=0}^{T-1} \|M_t\|_*^2.$$

Obviously, we can choose γ in an optimal way ("optimal" in the sense that we minimize the right-hand side of the above inequality) if a bound on $\sum_{t=0}^{T-1}\|M_t\|_*^2$ is known; see Section 3.2 for more details.

Alternatively, we might set $\gamma_t = -\ln(1-\gamma)$ and $\beta_t = 1$ for any $t \geq 0$. With this choice, Nesterov's Dual Averaging scheme still coincides with Algorithm 7.1 and the above bound on the regret remains valid.

7.2 Application in Semidefinite Optimization

Arora and Kale [AK07] discovered that a subfamily of (SDP) can be solved by using the Matrix Multiplicative Weights Update method with a carefully designed strategy for constructing the sequence $(M_t)_{t=0}^{T-1}$ as instrumental tool. Their method works provided that (SDP) complies with the following mild assumptions, which shall hold for the remainder of this chapter. Also, recall that Strong Duality still holds by Assumption 6.1.

Assumption 7.1 *In Formulation (SDP), we have $A_1 = \mathbb{I}_n/R$ for an $R > 0$ and $\mathrm{Tr}(C) > 0$.*

The first condition corresponds to a simple scaling constraint that ensures that the trace of any feasible matrix is bounded from above by R. This constraint is naturally present in the semidefinite relaxations of the combinatorial optimization problems that are considered in [AK07].

The second requirement is used to bound the optimal value of (SDP) by a positive quantity from below. Recall from Remark 6.1 that we have chosen $\alpha > 0$ such that $\mathrm{Tr}(A_j) < \alpha$ for any $1 \leq j \leq m$. As the matrix $\mathbb{I}_n/\alpha$

is feasible, the optimal value φ^* to (SDP) is thus bounded from below by $\mathrm{Tr}(C)/\alpha$. In particular, the optimal value φ^* is positive. On the other hand, we can bound φ^* from above by $\sum_{j=1}^{m} \bar{y}_j$, where $\bar{y}$ is given by Assumption 6.1.

The Multiplicative Weights Update method is a Dual Averaging scheme. Dual Averaging schemes are applicable to optimization problems that comply with Assumption 2.1. However, Problem (SDP) does not satisfy this requirement. In a first step, an appropriate problem transformation is thus required. Let us present the transformation used by Arora and Kale now.

7.2.1 Replacement by a cascade of feasibility problems

In order to find an approximation to the optimal value φ^* of (SDP), Arora and Kale perform a Binary Search over the interval $[\mathrm{Tr}(C)/\alpha, \sum_{j=1}^{m} \bar{y}_j]$. At every iteration of the Binary Search, they pose the following feasibility question.

Question 7.2 (Feasibility problem) *Given the current guess φ of φ^*, does there there exist a matrix $X \succeq 0$ with $\langle C, X \rangle_F > \varphi$ and such that $\langle A_j, X \rangle_F \leq 1$ for any $1 \leq j \leq m$?*

For the time being, let us assume that there exists an oracle which gives us approximate answers to the above question. More precisely, we suppose that there exists a **feasibility oracle** $\mathcal{F}$ with the following property. When we call $\mathcal{F}$ with input $(\varphi, \delta) \in [\mathrm{Tr}(C)/\alpha, \sum_{j=1}^{m} \bar{y}_j] \times (0, 1)$, it either returns a matrix $X \succeq 0$ such that:

$$\langle C, X \rangle_F > \varphi \qquad \text{and} \qquad \langle A_j, X \rangle_F \leq 1 \ \forall \ 1 \leq j \leq m,$$

or it declares correctly that there cannot exist any $X \succeq 0$ that satisfies both

$$\langle C, X \rangle_F > (1 + \delta)\varphi \qquad \text{and} \qquad \langle A_j, X \rangle_F \leq 1 \ \forall \ 1 \leq j \leq m.$$

The answers of this oracle are used in the Binary Search of Arora and Kale; see Algorithm 7.2. For this approximate Binary Search, we have the following convergence result.

Theorem 7.2 *[AK07] Assume that $\hat{X}$ is returned by Algorithm 7.2 after $T \in \mathbb{N}$ iterations. Then,*

$$\varphi^* - \left\langle C, \hat{X} \right\rangle_F \leq \left(\frac{5}{6}\right)^T \left(\sum_{j=1}^{m} \bar{y}_j - \mathrm{Tr}(C)/\alpha\right).$$

Algorithm 7.2 Approximate Binary Search [AK07]

1: Choose $T \in \mathbb{N}$.
2: Set $\hat{X} = \mathbb{I}_n/\alpha$, $l_0 = \mathrm{Tr}(C)/\alpha$, $u_0 = \sum_{j=1}^m \bar{y}_j$, and $\varphi_0 = (u_0 + l_0)/2$.
3: **for** $0 \leq t \leq T-1$ **do**
4:     Set $\delta_t = \frac{2(u_t - l_t)}{3(u_t + l_t)}$.
5:     Call feasibility oracle $\mathcal{F}$ with input (φ_t, δ_t).
6:     **if** $\mathcal{F}$ returns a feasible point X_{t+1} with $\langle C, X_{t+1} \rangle_F > \varphi_t$ **then**
7:         Update $\hat{X} = X_{t+1}$, $l_t = \langle C, X_{t+1} \rangle_F$, and $u_{t+1} = u_t$.
8:     **else**
9:         Define $l_{t+1} = l_t$ and $u_{t+1} = (1 + \delta_t)\varphi_t$.
10:     **end if**
11:     Set $\varphi_{t+1} = (u_{t+1} + l_{t+1})/2$.
12: **end for**
13: Return $\hat{X}$.

As the proof of this theorem is not given in the paper of Arora and Kale, we verify the correctness of this result in Appendix B.2. The above theorem shows that we need to perform about

$$T = \mathcal{O}\left(\ln \left[\frac{\sum_{j=1}^m \bar{y}_j}{\epsilon} \right] \right)$$

iterations of Algorithm 7.2 for obtaining a feasible ϵ-solution to Problem (SDP). We are left with the construction of the feasibility oracle $\mathcal{F}$.

7.2.2 Arora and Kale's implementation of the feasibility oracle

In [AK07], the feasibility oracle is realized as a particular instanciation of Algorithm 7.1 with a very specific strategy for constructing the matrices M_t. For $\varphi \in [\mathrm{Tr}(C)/\alpha, \sum_{j=1}^m \bar{y}_j]$, we define the set $\bar{\Delta}_m^\varphi$ as

$$\bar{\Delta}_m^\varphi := \left\{ y \in \mathbb{R}^m : y \geq 0, \ \sum_{j=1}^m y_j \leq \varphi \right\}.$$

In order to specify the matrices $(M_t)_{t=0}^{T-1}$ in Algorithm 7.1, Arora and Kale make use of a second oracle $\mathcal{G}$. When we call this oracle $\mathcal{G}$ with input matrix $X \in \Delta_n^M$, it either returns a vector $y \in \bar{\Delta}_m^\varphi$ with

$$\sum_{j=1}^m y_j \langle A_j, X \rangle_F - \langle C, X \rangle_F \geq 0, \tag{7.1}$$

or it declares correctly that there exists no vector $y \in \bar{\Delta}_m^\varphi$ satisfying (7.1). In addition, we assume that there exists a constant $\rho > 0$ such that

$$\left\| \sum_{j=1}^m A_j y_j - C \right\|_{(\infty)} \leq \rho$$

for all vectors $y \in \bar{\Delta}_m^\varphi$ that are returned by the oracle $\mathcal{G}$ for any input $X \in \Delta_n^M$. We refer to the minimal ρ that satisfies the above condition as the **width parameter** of the oracle $\mathcal{G}$. As we will see later on in this chapter, the smaller the parameter ρ is, the more attractive the running time of the resulting method becomes.

Remark 7.1 (Brute-force implementation of the oracle $\mathcal{G}$) *Choose a matrix $X \in \Delta_n^M$. The brute-force implementation of the oracle $\mathcal{G}$ looks as follows. We compute:*

$$j^* \in \arg \max_{1 \leq j \leq m} \langle A_j, X \rangle_F .$$

If $\varphi \langle A_{j^*}, X \rangle_F \geq \langle C, X \rangle_F$, we return φe_{j^*}, where e_{j^*} denotes the j^*-th unit vector in $\mathbb{R}^m$. Else, there cannot exist an $y \in \bar{\Delta}_m^\varphi$ satisfying (7.1). For this oracle, we obtain:

$$\rho = \max_{1 \leq j \leq m} \|\varphi A_j - C\|_{(\infty)} \leq \varphi \max_{1 \leq j \leq m} \|A_j\|_{(\infty)} + \|C\|_{(\infty)} .$$

Remark 7.2 (Arora and Kale's implementation of the oracle $\mathcal{G}$) *In their paper, Arora and Kale [AK07] consider several problems arising as semidefinite relaxations of NP-hard problems (for instance the semidefinite relaxation of MaxCut). For each of these problems, they define a particular oracle $\mathcal{G}$, which ideally has a smaller width parameter than the straightforward implementation presented above and that allows fast matrix exponential approximations. We postpone the discussion of the second issue to a later point in this section. However, we are not going to study the problem specific oracle constructions in this thesis, as they are not broadly applicable.*

Let us analyze the two situations that are described by the oracle outputs. We assume that we call the oracle $\mathcal{G}$ with the input matrix $X \in \Delta_n^M$.

Lemma 7.1 *[AK07] If there exists a vector $y \in \bar{\Delta}_m^\varphi$ that fulfills (7.1), then X is infeasible for (SDP) or $\langle C, X \rangle_F \leq \varphi$.*

Proof: Let $y \in \bar{\Delta}_m^\varphi$ fulfill Condition (7.1) for input matrix X. In addition, we suppose that X is feasible for (SDP) and that X satisfies $\langle C, X \rangle_F > \varphi$. Then,

$$\sum_{j=1}^m y_j \langle A_j, X \rangle_F - \langle C, X \rangle_F < \sum_{j=1}^m y_j - \varphi \leq \varphi - \varphi = 0,$$

which contradicts Condition (7.1). ∎

Note that the arguments of the proof above can be used to show that any positive scaling aX of the matrix X is infeasible for (SDP) or $\langle C, aX \rangle_F \leq \varphi$ if there is no $y \in \bar{\Delta}_m^\varphi$ satisfying (7.1).

Lemma 7.2 *[AK07] Assume that the oracle $\mathcal{G}$ declares that there is no element $y \in \bar{\Delta}_m^\varphi$ that satisfies (7.1) for the input matrix $X \in \Delta_n^M$. Then, the matrix*

$$\hat{X} := \frac{X}{\max_{1 \leq j \leq m} \langle A_j, X \rangle_F}$$

is a well-defined and feasible solution to (SDP) with an objective function value that is strictly larger than φ.

Proof: Assume that oracle $\mathcal{G}$ outputs that there is no $y \in \bar{\Delta}_m^\varphi$ that fulfills (7.1) for input matrix $X \in \Delta_n^M$. We immediately recognize that $\langle C, X \rangle_F > 0$, as otherwise the all zero vector would satisfy Condition (7.1). Applying Lemma 6.1, we obtain that the matrix

$$\hat{X} := \frac{X}{\max_{1 \leq j \leq m} \langle A_j, X \rangle_F}$$

is well-defined and positive semidefinite. In addition, the oracle answer guarantees that

$$\max_{y \in \bar{\Delta}_m^\varphi} \sum_{j=1}^{m} y_j \langle A_j, X \rangle_F < \langle C, X \rangle_F .$$

By Lemma 6.1, the left-hand side of the above inequality is strictly positive, which allows us to rewrite the inequality as

$$\varphi \max_{1 \leq j \leq m} \langle A_j, X \rangle_F < \langle C, X \rangle_F .$$

Dividing both sides by $\max_{1 \leq j \leq m} \langle A_j, X \rangle_F$, we immediately obtain that

$$\varphi < \left\langle C, \hat{X} \right\rangle_F .$$

Finally, we clearly have $\max_{1 \leq j \leq m} \left\langle A_j, \hat{X} \right\rangle_F = 1$. ∎

The outputs of the oracle $\mathcal{G}$ are used to construct the matrices M_t in Algorithm 7.3. If the oracle $\mathcal{G}$ returns an element $y_t \in \bar{\Delta}_m^\varphi$ for input matrix X_t at iteration $0 \leq t \leq T - 1$, we set:

$$M_t = \frac{1}{2\rho} \left(\sum_{j=1}^{m} A_j y_{t,j} - C + \rho \mathbb{I}_n \right) .$$

Algorithm 7.3 Arora and Kale's implementation of $\mathcal{F}$ [AK07]

1: Choose $T \in \mathbb{N}$ and an update parameter $0 < \gamma < 1$.
2: Let $X_0 = \mathbb{I}_n/n$ and $S_0 = 0 \in \mathcal{S}^n$.
3: **for** $0 \leq t \leq T - 1$ **do**
4:     Call the oracle $\mathcal{G}$ with input matrix X_t.
5:     **if** $\mathcal{G}$ returns a vector $y_t \in \bar{\Delta}_m^\varphi$ satisfying (7.1) **then**
6:         Set $M_t = \frac{1}{2\rho} \left(\sum_{j=1}^m A_j y_{t,j} - C + \rho \mathbb{I}_n \right)$ and $S_{t+1} = S_t + M_t$.
7:         Update $W_{t+1} = (1 - \gamma)^{S_{t+1}}$.
8:         Set $X_{t+1} = W_{t+1}/\mathrm{Tr}(W_{t+1})$.
9:     **else**
10:         Terminate and return

$$\hat{X} := \frac{X_t}{\max_{1 \leq j \leq m} \langle A_j, X_t \rangle_F}.$$

11:     **end if**
12: **end for**

By the definition of ρ, we have:

$$-\rho \leq \lambda_{\min} \left(\sum_{j=1}^m A_j y_{t,j} - C \right) \leq \lambda_{\max} \left(\sum_{j=1}^m A_j y_{t,j} - C \right) \leq \rho,$$

and thus $0 \preceq M_t \preceq \mathbb{I}_n$. In particular, M_t can be interpreted as the matrix chosen by the second player in the Matrix Multiplicative Weights Update method described in Algorithm 7.1. As an immediate consequence, we observe that Arora and Kale's Algorithm 7.3 is indeed a Matrix Multiplicative Weights Update method, but with one minor difference. While all T iterations are carried out in Algorithm 7.1, Algorithm 7.3 terminates earlier if the oracle $\mathcal{G}$ fails in finding an element $y \in \bar{\Delta}_m^\varphi$ that satisfies (7.1). If this case does not occur, we have the following guarantee.

Theorem 7.3 (Theorem 2 in [AK07]) *Let $\delta > 0$. We perform*

$$T \geq \frac{8 \ln(n) \rho^2 R^2}{\delta^2 \varphi^2}$$

iterations of Algorithm 7.3 with update parameter

$$\gamma = 1 - \exp \left(-\sqrt{\frac{2 \ln(n)}{T}} \right).$$

If the oracle $\mathcal{G}$ outputs at every iteration $0 \le t \le T - 1$ an element $y_t \in \bar{\Delta}_m^{\varphi}$ satisfying (7.1), the vector

$$\hat{y} = \delta \varphi e_1 + \frac{1}{T} \sum_{t=0}^{T-1} y_t$$

is feasible for the Dual Problem (6.2) and has an objective function value of at most $(1 + \delta)\varphi$, where e_1 denotes the first unit vector in $\mathbb{R}^m$.

Note that the update parameter γ in the above theorem deviates from the parameter that is used by Arora and Kale in their paper [AK07]. We end up with a different update parameter than Arora and Kale, because we utilize Theorem 7.1 in the proof below, and this theorem differs slightly from its analog in [AK07]. Although we use a different update parameter, we obtain the same bound on the iteration number T as Arora and Kale.

Proof of Theorem 7.3: Let $\delta > 0$. All the notations that we use in this proof refer to Algorithm 7.3. Assume that

$$T \ge \frac{8 \ln(n) \rho^2 R^2}{\delta^2 \varphi^2} \tag{7.2}$$

iterations of this algorithm are performed and that the oracle $\mathcal{G}$ returns at every iteration $0 \le t \le T - 1$ a vector $y_t \in \bar{\Delta}_m^{\varphi}$ that fulfills (7.1). Due to the definition of ρ, we obtain $\|M_t\|_{(\infty)}^2 \le 1$ for any $1 \le t \le T$. In addition, we have:

$$\sum_{t=0}^{T-1} \langle M_t, X_t \rangle_F = \frac{1}{2\rho} \sum_{t=0}^{T-1} \left\langle \sum_{j=1}^{m} A_j y_{t,j} - C, X_t \right\rangle_F + \frac{1}{2} \sum_{t=0}^{T-1} \operatorname{Tr}(X_t) \ge \frac{T}{2},$$

where the inequality holds because $X_t \in \Delta_n^M$ and as all y_t satisfy Condition (7.1). As Algorithm 7.3 is a Matrix Multiplicative Weights Update method, we may apply Theorem 7.1. This theorem yields to:

$$\min_{X \in \Delta_n^M} \left\{ \sum_{t=0}^{T-1} \langle M_t, X \rangle_F \right\} \ge \frac{\ln(n)}{\ln(1-\gamma)} + \frac{T(1 + \ln(1-\gamma))}{2},$$

that is,

$$\frac{1}{2\rho} \min_{X \in \Delta_n^M} \left\{ \sum_{t=0}^{T-1} \left\langle \sum_{j=1}^{m} A_j y_{t,j} - C, X \right\rangle_F \right\} \ge \frac{\ln(n)}{\ln(1-\gamma)} + \frac{T \ln(1-\gamma)}{2}.$$

For

$$\gamma = 1 - \exp\left(-\sqrt{\frac{2 \ln(n)}{T}}\right),$$

the above inequality can be rewritten as

$$\frac{1}{2\rho} \min_{X \in \Delta_n^M} \left\{ \sum_{t=0}^{T-1} \left\langle \sum_{j=1}^{m} A_j y_{t,j} - C, X \right\rangle_F \right\} \geq -\sqrt{2T \ln(n)},$$

which implies:

$$\lambda_{\min} \left(\frac{1}{T} \sum_{t=0}^{T-1} \sum_{j=1}^{m} A_j y_{t,j} - C \right) \geq -2\rho\sqrt{\frac{2\ln(n)}{T}}. \tag{7.3}$$

Let $\hat{y} = \delta\varphi e_1 + \frac{1}{T} \sum_{t=0}^{T-1} y_t$. As $y_t \in \bar{\Delta}_m^\varphi$ for any $0 \leq t \leq T - 1$, we clearly have:

$$\sum_{j=1}^{m} \hat{y}_j \leq (1+\delta)\varphi.$$

Exploiting Assumption 7.1 (that is, $A_1 = \mathbb{I}_n/R$), Inequality (7.3), and Condition (7.2), we finally obtain:

$$\begin{aligned}
\lambda_{\min} \left(\sum_{j=1}^{m} A_j \hat{y}_j - C \right) &= \frac{\delta\varphi}{R} + \lambda_{\min} \left(\frac{1}{T} \sum_{t=0}^{T-1} \sum_{j=1}^{m} A_j y_{t,j} - C \right) \\
&\geq \frac{\delta\varphi}{R} - 2\rho\sqrt{\frac{2\ln(n)}{T}} \\
&\geq 0,
\end{aligned}$$

which shows that $\hat{y}$ is feasible to (6.2). ∎

Arora and Kale utilize Algorithm 7.3 as feasibility oracle $\mathcal{F}$ in Algorithm 7.2. The resulting method has the following analytical complexity.

Corollary 7.3 *[AK07] Let $\epsilon > 0$. Running Algorithm 7.2 with Algorithm 7.3 as feasibility oracle $\mathcal{F}$, we need to perform*

$$\mathcal{O} \left(\frac{\alpha^2 \rho^2 R^2 \left[\sum_{j=1}^{m} \bar{y}_j \right]^2 \ln[n]}{\epsilon^2 \mathrm{Tr}^2[C]} \ln \left[\frac{\sum_{j=1}^{m} \bar{y}_j}{\epsilon} \right] \right)$$

iterations of Algorithm 7.3 in order to find a feasible ϵ-solution to (SDP).

Proof: Let $\epsilon > 0$. At every iteration t of Algorithm 7.2, we call Algorithm 7.3 with an accuracy of

$$\delta_t = \frac{2(u_t - l_t)}{3(u_t + l_t)} \geq \frac{\epsilon}{3 \sum_{j=1}^{m} \bar{y}_j},$$

where u_t and l_t are given by Algorithm 7.2. The concluding inequality holds as we can terminate Algorithm 7.2 as soon as $u_t - l_t \leq \epsilon$. It remains to multiply the analytical complexity of Algorithm 7.2 displayed in Theorem 7.2 with the number of iterations required by Algorithm 7.3; see Theorem 7.3. For the later result, we recall that any guess φ_t of φ^* in Algorithm 7.2 can be bounded from below by $\mathrm{Tr}(C)/\alpha$. ∎

It remains to analyze the cost per iteration. At every iteration $0 \leq t \leq T-1$ of Algorithm 7.3, the cost-dominating computations are as follows:

1. We must compute the matrix exponential

$$(1 - \gamma)^{S_{t+1}} = \exp\left(U_{t+1}\right), \qquad U_{t+1} := -\sqrt{\frac{2\ln(n)}{T}} S_{t+1}.$$

Using standard techniques, that is, computing the matrix exponential through an eigendecomposition of the symmetric matrix U_{t+1}, this operation requires $\mathcal{O}(n^3)$ arithmetic operations; see [ML03] for the classical survey on this topic. In their paper [AK07], Arora and Kale suggest to replace $\exp(U_{t+1})$ by a random approximation. Basically, they exploit the fact that $\exp(U_{t+1})$ can be written as the product $\exp(U_{t+1}/2)\exp(U_{t+1}/2)$. The rows of $\exp(U_{t+1}/2)$ are approximated by projecting an appropriate truncation of the matrix exponential Taylor series (about $\mathcal{O}(1/\epsilon)$ terms are considered) on $\mathcal{O}\left(1/\epsilon^2\right)$ random directions. As the resulting algorithm requires about $\mathcal{O}(\ln[1/\epsilon]/\epsilon^5)$ arithmetic operations to find a feasible ϵ-solution to Problem (SDP), we do not further elaborate on this random method. In particular, proving this result would go beyond the scope of this thesis. Note that we neglect all other problem parameters such as m and n in the previous complexity estimates.

2. We need $\mathcal{O}(mn^2)$ arithmetic operations for the computation of $\sum_{j=1}^{m} A_j y_{t,j}$.

3. Also, we need to consider the cost of a single call of the oracle $\mathcal{G}$. Implementing $\mathcal{G}$ as described in Remark 7.1, about $\mathcal{O}(mn^2)$ arithmetic operations are required per oracle call.

Provided that we utilize for $\mathcal{G}$ the straightforward implementation described in Remark 7.1, we need in total $\mathcal{O}(n^3 + mn^2)$ arithmetic operations per iteration. If all matrices A_j have at most S non-zero entries, this complexity estimate reduces to $\mathcal{O}(n^3 + mS)$. The following result summarizes the complexity study that we performed in this section.

Corollary 7.4 *[AK07] Assume that we use the Dual Averaging* (DA) *scheme described in Algorithm 7.3 as feasibility oracle $\mathcal{F}$ in the Binary Search* (BS)

Algorithm 7.2. Let the oracle $\mathcal{G}$ in Algorithm 7.3 be implemented as described in Remark 7.1. With

$$\nu := \frac{\epsilon}{R\left(\max_{1\leq k\leq m}\|A_k\|_{(\infty)} + \alpha\,\|C\|_{(\infty)}\,/\mathrm{Tr}(C)\right)\sum_{j=1}^{m}\bar{y}_j},$$

the resulting method requires

$$\mathrm{Compl}_{\mathrm{BS+DA}}(\mathrm{SDP},\epsilon) = \mathcal{O}\left(\frac{[n^3 + mS]\ln[n]}{\nu^2}\ln\left[\frac{\sum_{j=1}^{m}\bar{y}_j}{\epsilon}\right]\right)$$

arithmetic operations to find a feasible ϵ-solution to (SDP).

If we consider only the parameters m, n, and ϵ, the above complexity results answer affirmatively Question (7.1). However, the quality of this answer is very rudimentary. First, the above complexity result depends on many more problem parameters than just ϵ, m, and n. Namely, it is also affected by R, the initial lower and upper bound on the optimal value of (SDP), and the norms of A_j and C. And second, the high iteration cost due to the matrix exponentiation and the comparatively strong influence of $\epsilon > 0$ give little hope to find a meaningful and reasonably large subclass of Problems (SDP) with $n \geq m$, where we can theoretically beat Interior-Point methods. Let us be more precise. Suppose that the data of (SDP) is chosen so that $m \leq n$ and such that $mS = n^\beta$ with $\beta \in [2,3]$. These assumptions do no exclude the matrix $\sum_{j=1}^{m} A_j y_j$ from being fully dense. With ν being defined as in the above theorem, we have $\mathrm{Compl}_{\mathrm{BS+DA}}(\mathrm{SDP},\epsilon) \ll \mathrm{Compl}_{\mathrm{IP}}(\mathrm{SDP},\epsilon)$ if and only if

$$1/\nu^2 \ll m\sqrt{n},$$

where we neglect all logarithmic terms. This condition is very restrictive, as it already fails for $\nu \leq n^{-3/4}$. However, if $n \ll m$, the setting is fully in favor of Arora and Kale's method, as the arithmetic complexity of their method grows only linearly in m. In contrast, the efficiency estimate of Interior-Point methods grows faster than cubically in m. In future, we will thus always focus on the more critical situation where $n \geq m$.

In the next chapter, we will see that we can apply Dual Averaging schemes in an alternative way and obtain a better worst-case complexity result due to lower iteration cost.

Chapter 8

From semidefinite optimization problems to matrix saddle-point problems

IN this and the subsequent two chapters, we approach Question 7.1 by using Primal-Dual Subgradient methods, Smoothing Techniques, and Mirror-Prox methods. However, these methods cannot be applied to the current formulation of Problem (SDP). In this chapter, we show that we can express Problem (SDP) alternatively as a (cascade of) **matrix saddle-point problem**(s), which meet(s) both Assumption 2.1 and the structural requirements of Smoothing Techniques and Mirror-Prox methods.

In Section 8.1, we reuse the problem transformation utilized by Arora and Kale in [AK07]. In contrast to them, we answer the feasibility problem described in Question 7.2 by solving a matrix saddle-point problem approximately. In Section 8.2, we present an alternative problem transformation that also yields to a matrix saddle-point reformulation of Problem (SDP). While the first transformation based on Arora and Kale's strategy requires Assumption 7.1 to hold, this alternative approach asks for a positive definite matrix C in (SDP).

Matrix saddle-point problems can be described in a primal and in a dual form. As we observe in Section 8.3, where we apply Primal-Dual Subgradient methods (more specifically Mirror-Descent schemes) to both the primal and the dual version of the matrix saddle-point problem, the dual description has significant computational advantages over its primal counterpart. More precisely, when applying Mirror-Descent methods to the primal problem version, we are supposed to carry out a matrix exponentiation at every iteration.

However, with the dual strategy, we only need to derive the maximal eigenvalue and the leading eigenvector of a symmetric matrix at every iteration, yielding to a lower iteration cost. Combining the resulting complexity estimate with the problem transformations studied in Section 8.1 and 8.2, we get an affirmative answer to Question 7.1, that is, we obtain an algorithm that is fast enough to meet our complexity requirements. More importantly, this answer is much less restrictive than the one presented in Corollary 7.4. Finally, we conclude this chapter by applying Interior-Point methods to matrix saddle-point problems in Section 8.4. We compare their efficiency estimates with the complexity result of Mirror-Descent methods and show that there exist a meaningful group of large-scale matrix saddle-point problems where Mirror-Descent methods outperform Interior-Point methods theoretically.

Contributions: The transformation of (SDP) into a matrix saddle-point problem that we present in Section 8.2 is original. The first problem transformation is known.

Relevant literature: We use the reference [NY83]. Parts of this chapter are taken from [BB09, BBN11].

8.1 Binary Search based on matrix saddle-point problems

In this section, we show that we can answer the Feasibility Question 7.2 by approximately solving an appropriate matrix saddle-point problem. More importantly, the approximate solution to this matrix saddle-point problem can be utilized to implement the feasibility oracle $\mathcal{F}$ that is described in Section 7.2.1. As an immediate consequence, we recognize that the optimal value of Problem (SDP) can be approximated by solving a cascade of matrix saddle-point problems. Let us introduce these matrix saddle-point problems now.

We use the same notation as in the previous chapter. As in the setting of Arora and Kale [AK07], we impose the following assumptions:

Assumption 8.1 *The trace of C is positive and $A_1 = \mathbb{I}_n/R$ for a constant $R > 0$.*

We choose $\mathrm{Tr}(C)/\alpha \leq \varphi \leq \sum_{j=1}^{m} \bar{y}_j$ and consider the matrix saddle-point problem:

$$\min_{X \in \Delta_n^M} \max_{y \in \Delta_m} \sum_{j=1}^{m} y_j \langle A_j - C/\varphi, X \rangle_F = \min_{X \in \Delta_n^M} \max_{1 \leq j \leq m} \langle A_j - C/\varphi, X \rangle_F . \quad (8.1)$$

First of all, Problem (8.1) is in a form such that Primal-Dual Subgradient methods, Smoothing Techniques, and Mirror-Prox methods are applicable to it. On the one hand, Problem (8.1) satisfies Assumption 2.1. For both the standard simplex Δ_m and the matrix simplex Δ_n^M, we know distance-generating functions for which the optimal solution to Problem (2.10) can be written in a closed form; see Examples 2.14 and 2.15. On the other hand, Problem (8.1) complies with the structural requirements specified in (4.1). When we apply one of the above methods to Problem (8.1), we obtain an element $\bar{X} \in \Delta_n^M$ with the following guarantee:

$$\max_{1 \leq j \leq m} \left\langle A_j - C/\varphi, \bar{X} \right\rangle_F - \min_{X \in \Delta_n^M} \max_{1 \leq j \leq m} \left\langle A_j - C/\varphi, X \right\rangle_F \leq \bar{\epsilon}, \qquad (8.2)$$

where $\bar{\epsilon} > 0$ is the solution accuracy. Assume now that $\bar{\epsilon} = \delta/R$ for some $\delta > 0$. Clearly, the method output $\bar{X}$ satisfies exactly one of the following two conditions:

1. either we have $\max_{1 \leq j \leq m} \left\langle A_j - C/\varphi, \bar{X} \right\rangle_F < 0$;

2. or $\max_{1 \leq j \leq m} \left\langle A_j - C/\varphi, \bar{X} \right\rangle_F \geq 0$, which implies by (8.2) that

$$\min_{X \in \Delta_n^M} \max_{1 \leq j \leq m} \left\langle A_j - C/\varphi, \bar{X} \right\rangle_F \geq -\delta/R.$$

The output $\bar{X}$ can be used to construct a feasibility oracle $\mathcal{F}$ as defined in Section 7.2.1. This procedure is based on the two next lemmas.

Lemma 8.1 *Let $\bar{X} \in \Delta_n^M$ satisfy $\max_{1 \leq j \leq m} \left\langle A_j - C/\varphi, \bar{X} \right\rangle_F < 0$. The matrix $\hat{X} := \bar{X}/\max_{1 \leq j \leq m} \left\langle A_j, \bar{X} \right\rangle_F$ is well-defined and constitutes a feasible solution to (SDP) with an optimal value that is strictly larger than φ.*

Proof: Let $\bar{X} \in \Delta_n^M$ such that $\max_{1 \leq j \leq m} \left\langle A_j - C/\varphi, \bar{X} \right\rangle_F < 0$. If $\left\langle C, \bar{X} \right\rangle_F \leq 0$, then

$$1/R \leq \max_{1 \leq j \leq m} \left\langle A_j, \bar{X} \right\rangle_F \leq \max_{1 \leq j \leq m} \left\langle A_j - C/\varphi, \bar{X} \right\rangle_F < 0,$$

which is a contradiction. We obtain:

$$\max_{y \in \bar{\Delta}_m^\varphi} \sum_{j=1}^m y_j \left\langle A_j, \bar{X} \right\rangle_F - \left\langle C, \bar{X} \right\rangle_F < 0.$$

It remains to apply the same arguments as in the proof of Lemma 7.2. ∎
We have the following guarantee for the second case.

Lemma 8.2 *If*

$$\min_{X \in \Delta_n^M} \max_{1 \leq j \leq m} \langle A_j - C/\varphi, X \rangle_F \geq -\delta/R, \tag{8.3}$$

there exists no feasible point to (SDP) with an objective function value that is strictly larger than $(1 + \delta)\varphi$.

Proof: Assume that (8.3) holds. Then, there exists an element $\hat{y} \in \Delta_m$ such that:

$$\lambda_{\min}\left(\varphi \sum_{j=1}^{m} A_j \hat{y}_j - C\right) \geq -\frac{\delta\varphi}{R}.$$

In particular, it holds that:

$$-\frac{\delta\varphi}{R} + \lambda_{\min}\left(\frac{(\hat{y}_1 + \delta)\mathbb{I}_n\varphi}{R} + \varphi \sum_{j=2}^{m} A_j \hat{y}_j - C\right) \geq -\frac{\delta\varphi}{R},$$

which shows that $y := \varphi(\hat{y} + \delta e_1)$ is a feasible solution to (6.2), where e_1 denotes the first unit vector in $\mathbb{R}^m$. We conclude by recalling that by definition the components of $\hat{y}$ sum up to one. ∎

In order to find a feasible ϵ-solution to Problem (SDP), we thus need to solve approximately about

$$\mathcal{O}\left(\ln\left[\frac{\sum_{j=1}^{m} \bar{y}_j}{\epsilon}\right]\right)$$

Matrix Saddle-Point Problems (8.1). Each of these auxiliary problems needs to be solved up to an accuracy of

$$\bar{\epsilon} = \frac{\delta}{R} \geq \frac{\epsilon}{3R \sum_{j=1}^{m} \bar{y}_j}; \tag{8.4}$$

see Algorithm 7.2 and Theorem 7.2 for the details.

8.2 An alternative problem transformation

In this section, we give an alternative matrix saddle-point problem reformulation of (SDP). While the previous transformation requires a scaling constraint as described in Assumption 8.1, this alternative approach asks for a positive definite matrix C in (SDP). For reference, let us formalize this requirement:

Assumption 8.2 *The matrix C is positive definite.*

Let this assumption hold for this section. As a consequence, we obtain that the inequality $\max_{1\leq j\leq m} \langle A_j, X\rangle_F > 0$ holds for every non-vanishing $X \succeq 0$. Indeed, we have $\langle C, X\rangle_F > 0$ for every non-zero $X \succeq 0$, as $C \succ 0$. It remains to apply Lemma 6.1.

The following lemma constitutes the first step in the reformulation of Problem (SDP).

Lemma 8.3 *We have:*

$$\max_{X\succeq 0}\left\{\langle C, X\rangle_F : \max_{1\leq j\leq m}\langle A_j, X\rangle_F \leq 1\right\} = \left(\min_{Y\in Q'}\max_{1\leq j\leq m}\langle A_j, Y\rangle_F\right)^{-1}, \quad (8.5)$$

where $Q' := \left\{Y \succeq 0 : \langle C, Y\rangle_F = 1\right\}$.

Proof: Let us first prove the equality $\varphi^* = \varphi'$, where

$$\varphi^* := \max_{X\succeq 0}\left\{\langle C, X\rangle_F : \max_{1\leq j\leq m}\langle A_j, X\rangle_F \leq 1\right\},$$

$$\varphi' := \sup_{X\succeq 0}\frac{\langle C, X\rangle_F}{\max_{1\leq j\leq m}\langle A_j, X\rangle_F}.$$

The function $X \mapsto \langle C, X\rangle_F / \max_{1\leq j\leq m}\langle A_j, X\rangle_F$ is homogeneous in X and strictly positive for all non-zero $X \succeq 0$. This implies that we can compute its supremum over all $X \succeq 0$ by restricting ourselves to all matrices $X \in \Delta_n^M$. For any $X \in \Delta_n^M$, the ratio $\langle C, X\rangle_F / \max_{1\leq j\leq m}\langle A_j, X\rangle_F$ lies in $(0, \infty)$, because both $\langle C, X\rangle_F$ and $\max_{1\leq j\leq m}\langle A_j, X\rangle_F$ belong to the interval $(0, \infty)$. As $X \mapsto \langle C, X\rangle_F / \max_{1\leq j\leq m}\langle A_j, X\rangle_F$ is continuous on the compact set Δ_n^M, its maximum on this set is attained. In particular, the value φ' is finite and we can replace the "sup"- by a "max"-operator in the definition of φ'. Obviously, we have $\varphi^* \leq \varphi'$. Indeed, if $\bar{X}$ is an optimal solution to Problem (SDP), we have:

$$\max_{1\leq j\leq m}\langle A_j, \bar{X}\rangle_F = 1,$$

which holds due to the linearity of the objective function. Hence,

$$\langle C, \bar{X}\rangle_F = \frac{\langle C, \bar{X}\rangle_F}{\max_{1\leq j\leq m}\langle A_j, \bar{X}\rangle_F} \leq \varphi'.$$

On the other hand, we can choose a matrix $X' \neq 0$ such that

$$\varphi' = \frac{\langle C, X'\rangle_F}{\max_{1\leq j\leq m}\langle A_j, X'\rangle_F}.$$

Let $\tilde{X} := X'/\max_{1\leq j\leq m}\langle A_j, X'\rangle_F$. Evidently, $\tilde{X}$ is feasible for (SDP), and $\varphi' = \langle C, \tilde{X}\rangle_F \leq \varphi^*$. We conclude that $\varphi^* = \varphi'$.

Observe that Y belongs to Q' if and only if there exists $X \succeq 0$ such that $Y = X/\langle C, X \rangle_F$. This implies that we can write

$$\max_{X \succeq 0} \left\{ \frac{\langle C, X \rangle_F}{\max_{1 \leq j \leq m} \langle A_j, X \rangle_F} \right\} = \max_{Y \in Q'} \left(\max_{1 \leq j \leq m} \langle A_j, Y \rangle_F \right)^{-1}.$$

This last problem is equivalent to the right-hand side of (8.5). ∎

We have converted our original problem into the following auxiliary format:

$$\min_{Y \in Q'} \max_{1 \leq j \leq m} \langle A_j, Y \rangle_F = \min_{Y \in Q'} \max_{y \in \Delta_m} \sum_{j=1}^{m} y_j \langle A_j, Y \rangle_F,$$

where we use the notation of Lemma 8.3. Unfortunately, no distance-generating function is known for which we can give a closed form optimal solution to (2.10) with Q' as feasible set. The additional transformation step that is necessary to resolve this difficulty requires a Cholesky decomposition of the positive definite matrix C. Let $C = LL^T$, where $L \in \mathcal{M}^n$ is lower triangular with strictly positive diagonal entries. About $\mathcal{O}(n^3)$ arithmetic operations are needed to compute such a matrix L.

Let $Y \in Q'$ and define $\bar{Y} := L^T Y L \succeq 0$. As L is invertible, we can write:

$$\langle C, Y \rangle_F = \left\langle LL^T, L^{-T} \bar{Y} L^{-1} \right\rangle_F = \mathrm{Tr}(\bar{Y}),$$

and:

$$\langle A_j, Y \rangle_F = \left\langle A_j, L^{-T} \bar{Y} L^{-1} \right\rangle_F = \left\langle L^{-1} A_j L^{-T}, \bar{Y} \right\rangle_F \qquad \forall\, 1 \leq j \leq m.$$

We obtain:

$$\min_{Y \in Q'} \max_{1 \leq j \leq m} \langle A_j, Y \rangle_F = \min_{\bar{Y} \in \Delta_n^M} \max_{1 \leq j \leq m} \left\langle L^{-1} A_j L^{-T}, \bar{Y} \right\rangle_F.$$

Summarizing, we have proven the following lemma.

Proposition 8.4 *Let $B_j := L^{-1} A_j L^{-T}$ for any $1 \leq j \leq m$. We have:*

$$\varphi^* = \max_{X \succeq 0} \left\{ \langle C, X \rangle_F : \max_{1 \leq j \leq m} \langle A_j, X \rangle_F \leq 1 \right\} = \left(\min_{X \in \Delta_n^M} \max_{1 \leq j \leq m} \langle B_j, X \rangle_F \right)^{-1}.$$

Therefore, the new task consists in solving the following matrix saddle-point problem:

$$\min_{X \in \Delta_n^M} \max_{1 \leq j \leq m} \langle B_j, X \rangle_F = \min_{X \in \Delta_n^M} \max_{y \in \Delta_m} \sum_{j=1}^{m} y_j \langle B_j, X \rangle_F, \qquad (8.6)$$

where the matrices B_j, $1 \leq j \leq m$, are defined in Proposition 8.4. Note that this last transformation step can affect the possible structure of the original problem. For instance, the matrices B_j might be dense even though the matrices A_j can be sparse.

When we apply Primal-Dual Subgradient methods, Smoothing Techniques, or Mirror-Prox methods to Problem (8.6), we end up with an approximate solution to this problem. The next proposition shows how we can transform this method output in an approximate solution to the Initial Problem (SDP).

Proposition 8.5 *Let $\delta > 0$ and $\bar{X}$ be a feasible δ-solution to (8.6). The matrix*

$$\hat{X} := \frac{L^{-T}\bar{X}L^{-1}}{\max_{1 \leq j \leq m} \langle B_j, \bar{X} \rangle_F} \tag{8.7}$$

is well-defined and a feasible solution to Problem (SDP) with

$$\varphi^* - \left\langle C, \hat{X} \right\rangle_F \leq \frac{\delta \varphi^*}{\max_{1 \leq j \leq m} \langle B_j, \bar{X} \rangle_F}. \tag{8.8}$$

Proof: Let $\delta > 0$ and assume that $\bar{X}$ constitutes a feasible δ-solution to (8.6). As $\bar{X} \succeq 0$, we have $L^{-T}\bar{X}L^{-1} \succeq 0$ and $\max_{1 \leq j \leq m} \langle A_j, L^{-T}\bar{X}L^{-1} \rangle_F > 0$; see the short note below Assumption 8.2. Thus, the matrix $\hat{X}$ defined in (8.7) is well-defined and positive semidefinite. In addition, it holds that $\max_{1 \leq j \leq m} \langle A_j, \hat{X} \rangle_F = 1$, which shows that $\hat{X}$ is feasible for (SDP). We can easily see that:

$$\left\langle C, \hat{X} \right\rangle_F \max_{1 \leq j \leq m} \langle B_j, \bar{X} \rangle_F = \mathrm{Tr}(\bar{X}) = 1. \tag{8.9}$$

For the sake of notational simplicity, let us define the function:

$$h : \mathcal{S}^n \to \mathbb{R} : X \mapsto h(X) := \max_{1 \leq j \leq m} \langle B_j, X \rangle_F.$$

That is in fact the objective function of Matrix Saddle-Point Problem (8.6), and we write h^* for its minimal value on Δ_n^M. Since $\bar{X}$ is an δ-solution to (8.6), it holds that

$$h(\bar{X}) - h^* \leq \delta.$$

Thus, by Proposition 8.4 and Equation (8.9), we obtain:

$$\varphi^* - \left\langle C, \hat{X} \right\rangle_F = \frac{1}{h^*} - \frac{1}{h(\bar{X})} = \frac{h(\bar{X}) - h^*}{h^* h(\bar{X})} \leq \frac{\delta}{h^* h(\bar{X})} = \frac{\delta \varphi^*}{h(\bar{X})}.$$

$\blacksquare$

Let $\epsilon \in (0,1)$ and choose δ in the above proposition such that $\delta = \epsilon / \sum_{j=1}^{m} \bar{y}_j$. Then, the matrix $\hat{X}$ from (8.7) is a feasible solution to Semidefinite Optimization Problem (SDP) with a relative accuracy of at most ϵ. Indeed, we obtain by (8.8):

$$\varphi^* - \left\langle C, \hat{X} \right\rangle_F \leq \frac{\delta \varphi^*}{\max_{1 \leq j \leq m} \left\langle B_j, \bar{X} \right\rangle_F} = \delta \varphi^* \left\langle C, \hat{X} \right\rangle_F \leq \delta \varphi^* \sum_{j=1}^{m} \bar{y}_j = \epsilon \varphi^*.$$

The first equality comes from (8.9) and the last inequality holds as $\hat{X}$ is feasible for (SDP).

Let us summarize this subsection. We can find a feasible solution with relative accuracy $\epsilon \in (0,1)$ to (SDP) by solving approximately Auxiliary Problem (8.6). This auxiliary problem needs to be solved up to an accuracy of

$$\delta = \frac{\epsilon}{\sum_{j=1}^{m} \bar{y}_j}. \tag{8.10}$$

The forward transformation of the problem and the backward substitution of the approximate solution require in total $\mathcal{O}\left(mn^3\right)$ arithmetic operations.

8.3 Solving the primal or the dual problem?

In the two previous sections, we observed that we can compute an approximate solution to (SDP) by solving a (cascade of) matrix saddle-point problem(s), provided that Problem (SDP) complies with one of the Assumptions 8.1 or 8.2. Thus, our **core problem** is henceforth the following matrix saddle-point problem:

$$\min_{X \in \Delta_n^M} \overline{\psi}(X), \qquad \overline{\psi}(X) := \max_{y \in \Delta_m} - \sum_{j=1}^{m} y_j \left\langle D_j, X \right\rangle_F, \tag{PSP}$$

where $D_1, \ldots, D_n$ are symmetric $(n \times n)$-matrices. Note that we have the correspondence $D_j = -A_j + C/\varphi$ in Formulation (8.1), as the entries of the dual variable y sum up to one, and that $D_j = -B_j$ in (8.6) for any $1 \leq j \leq m$. We assume that every D_j has at most S non-zero entries and that $mS \geq n^2$, implying that the matrices $\sum_{j=1}^{m} y_j D_j$ can be fully dense. In addition, we define:

$$\mathcal{L} := \max_{1 \leq j \leq m} \|D_j\|_{(\infty)}.$$

According to the standard MiniMax Theorem in Convex Optimization (Corollary 37.3.2 in [Roc70]), we have the following correspondence:

$$\min_{X \in \Delta_n^M} \max_{y \in \Delta_m} - \sum_{j=1}^{m} y_j \left\langle D_j, X \right\rangle_F = \max_{y \in \Delta_m} \min_{X \in \Delta_n^M} - \sum_{j=1}^{m} y_j \left\langle D_j, X \right\rangle_F$$

$$= -\min_{y \in \Delta_m} \max_{X \in \Delta_n^M} \sum_{j=1}^{m} y_j \langle D_j, X \rangle_F.$$

We observe that both the primal problem and its dual counterpart are solvable by Primal-Dual Subgradient methods, Smoothing Techniques, and Mirror-Prox methods. However, from a complexity point of view, there are significant differences whether we solve the primal or the dual version of this problem, as Primal-Dual Subgradient methods and Smoothing Techniques make different computational demands on the inner and the outer problem. In this section, we apply Mirror-Descent methods – a subclass of Primal-Dual Subgradient methods – to both problems in order to point out the discrepancy of these different computational demands regarding the final arithmetic complexity of the algorithm.

8.3.1 Mirror-Descent schemes applied to the primal problem

At first, we run Mirror-Descent methods on (PSP).

We equip $\mathcal{S}^n$ with the induced 1-norm $\|\cdot\|_{(1)}$, for which $\|\cdot\|_{(\infty)}$ is the corresponding dual norm. In order to apply to Mirror-Descent methods, we choose $d_{\Delta_n^M}$ as distance-generating function on Δ_n^M. As Example 2.15 shows, this setting is in full accordance with Assumption 2.1.

Let us discuss now the iteration cost. At every iteration $t \in \mathbb{N}_0$ of Mirror-Descent methods, we need to compute a subgradient G_t of $\overline{\psi}$ at a point $X_t \in \text{relint}(\Delta_n^M)$. In view of Lemma 3.1.10 in [Nes04], we compute G_t as follows:

$$G_t := -D_{j^*}, \qquad j^* \in \arg \max_{1 \leq j \leq m} -\langle D_j, X_t \rangle_F.$$

This computation requires $\mathcal{O}(mS)$ arithmetic operations. In addition, the norm of all subgradients is bounded from above by the maximal norm of the matrices D_j:

$$\|G_t\|_{(\infty)} \leq \mathcal{L} \qquad \forall\, G_t \in \partial\overline{\psi}(X_t),\ \forall\, X_t \in \text{relint}(\Delta_n^M).$$

The computation of the projection $\pi_{\Delta_n^M, 1}(S_{t+1})$, $S_{t+1} \in \mathcal{S}^n$, is the other cost-dominating operation of Mirror-Descent methods; see Algorithm 3.1. We recall from Example 2.15 that the computation of this projection corresponds to carrying out a matrix exponentiation, which requires $\mathcal{O}(n^3)$ arithmetic operations.

Theorem 8.1 (Consequence of Chapter 3 in [NY83]) *Choose $\epsilon > 0$. Mirror-Descent* (MD) *methods require about*

$$\text{Compl}_{\text{MD}}(\text{PSP}, \epsilon) = \mathcal{O}\left(\frac{[n^3 + mS]\,\mathcal{L}^2 \ln[n]}{\epsilon^2}\right)$$

arithmetic operations to find a feasible ϵ-solution to Problem (PSP).

Proof: It is an immediate consequence of Theorem 3.2, Inequality (3.3), and Example 2.18. ∎

8.3.2 Mirror-Descent methods applied to the dual problem

Let us now apply Mirror-Descent schemes to the dual counterpart of (PSP), that is, to:

$$\min_{y \in \Delta_m} \overline{\phi}(y), \qquad \overline{\phi}(y) := \max_{X \in \Delta_n^M} \sum_{j=1}^{m} y_j \langle D_j, X \rangle_F = \lambda_{\max}\left(\sum_{j=1}^{m} y_j D_j \right). \qquad \text{(DSP)}$$

Note that we discard the minus sign in front of the problem. We consider $\mathbb{R}^m$ together with the 1-norm $\|\cdot\|_1$ and choose d_{Δ_m} as distance-generating function on the simplex Δ_m. As shown in Example 2.14, the evaluation of the parametrized mirror-operator $\pi_{\Delta_m,1}(s_{t+1})$ in Algorithm 3.2 requires $\mathcal{O}(m)$ arithmetic operations. In the current setting, the computation of the First-Order oracle is the operation that dictates the iteration cost. For any $y_t \in \mathrm{relint}(\Delta_m)$, we compute $g_t \in \partial \overline{\phi}(y)$ as follows:

$$g_t = \left(\langle D_j, X^* \rangle_F \right)_{j=1}^m, \qquad X^* \in \arg \min_{X \in \Delta_n^M} \sum_{j=1}^{m} y_{t,j} \langle D_j, X \rangle_F.$$

Let $\bar{D} := \sum_{j=1}^{m} y_{t,j} D_j$, which can be computed in $\mathcal{O}(mS)$ arithmetic operations. Given an eigendecomposition $Q(\bar{D})\mathrm{Diag}(\lambda(\bar{D}))Q^T(\bar{D})$ of $\bar{D}$ with $Q(\bar{D}) = [q_1(\bar{D}), \dots, q_n(\bar{D})]$, the matrix X^* can be written as

$$X^* = \lambda_{\max}(\bar{D})q_1(\bar{D})q_1^T(\bar{D}).$$

In other words, the matrix X^* is fully determined by the maximal eigenvalue and the leading of eigenvector of $\bar{D}$. Clearly, we can compute these entities through an eigendecomposition of the matrix $\bar{D}$, which requires $\mathcal{O}(n^3)$ arithmetic operations. However, we can also consider to derive them by running an (adapted) Power method, which gives us finally not the exact value of X^*, but an approximation to it. The effects of the use of approximate first-order information in Mirror-Descent schemes were analyzed by d'Aspremont in [d'A11]. Clearly, the use of inexact first-order information only makes sense when we get some computational advantage. According to [KW92], the complexity of computing the largest eigenvalue and its corresponding eigenvector of a symmetric matrix costs on average $\tilde{\mathcal{O}}(n^2 \sqrt{\mathcal{L}/\epsilon})$ arithmetic operations, where we use the $\tilde{\mathcal{O}}$-notation to indicate that all logarithmic terms are suppressed in this result. In our implementation, we have used a variant of the Power method designed by Arkadi Nemirovski that also runs in $\tilde{\mathcal{O}}(n^2\sqrt{\mathcal{L}/\epsilon})$ arithmetic operations on average. To the best of our knowledge, this variant has not been published so far.

As these methods produce stochastic approximations of both the maximal eigenvalue and the leading eigenvector, we refer to Mirror-Descent schemes with these approximation methods as **randomized Mirror-Descent methods**.

Theorem 8.2 (Consequence of Chapter 3 in [NY83]) *Randomized Mirror-Descent* (RMD) *methods require*

$$\mathrm{Compl}_{\mathsf{RMD}}^{\mathrm{rand}}(\mathrm{DSP}, \epsilon) = \tilde{\mathcal{O}}\left(\frac{\mathcal{L}^{2.5}n^2}{\epsilon^{2.5}} + \frac{\mathcal{L}^2 mS}{\epsilon^2}\right)$$

arithmetic operations to find an element $y \in \Delta_m$ for which the expectation of $\overline{\phi}(y)$ does not deviate more than ϵ from the optimal value to (DSP).

Note that we add a "rand"-tag to the complexity notion in order to indicate that the method involves some randomized parts. Of particular importance in our context, the method outputs not only an element $y \in \Delta_m$, but it gives also access to a matrix $X \in \Delta_n^M$ for which the difference $\overline{\phi}(y) - \min_{1 \le j \le m}\langle D_j, X\rangle_F$ is at most ϵ on average; we refer to Section 4 in [Nes09] for the construction of the matrix X, although the method that is studied there is non-randomized.

Comparison: We compare the efficiencies $\mathrm{Compl}_{\mathsf{RMD}}^{\mathrm{rand}}(\mathrm{DSP}, \epsilon)$ and $\mathrm{Compl}_{\mathsf{MD}}(\mathrm{PSP}, \epsilon)$. We write $mS = n^{\beta}$ with a $\beta \in [2, 3]$. Neglecting all logarithmic terms, we have $\mathrm{Compl}_{\mathsf{RMD}}^{\mathrm{rand}}(\mathrm{DSP}, \epsilon) \ll \mathrm{Compl}_{\mathsf{MD}}(\mathrm{PSP}, \epsilon)$ if

$$\max\left\{\sqrt{\mathcal{L}/\epsilon}, n^{\beta-2}\right\} \ll n.$$

We observe that it is complexity-wise more appealing to solve the Dual Formulation (DSP) instead of the Primal Version (PSP) if the matrices D_j are not dense (that is, $\beta < 3$) and if the ratio $\mathcal{L}/\epsilon$ is smaller than n^2. This later requirement is typically fulfilled if the problem is of large scale and needs to be solved up to a moderate accuracy. In particular, all conditions match very well the real-life setting that motivates Question 7.1. For all future comparisons of efficiency estimates, we thus use the arithmetic complexity $\mathrm{Compl}_{\mathsf{RMD}}^{\mathrm{rand}}(\mathrm{DSP}, \epsilon)$ of randomized Mirror-Descent methods.

Arithmetic complexity for (SDP): Finally, Theorem 8.2 yields the following corollaries. We recall that we neglect all logarithmic terms in the $\tilde{\mathcal{O}}$-notation.

Corollary 8.6 *Suppose that Assumption 8.1 holds and choose $\epsilon > 0$. We use the Binary Search (BS) Algorithm 7.2 to find a feasible ϵ-solution to Problem (SDP), where we apply at every iteration a randomized Mirror-Descent method to Problem (DSP). Defining*

$$\nu := \frac{\epsilon}{R\left(\max_{1\leq k\leq m}\|A_k\|_{(\infty)} + \alpha\|C\|_{(\infty)}/\mathrm{Tr}(C)\right)\sum_{j=1}^{m}\bar{y}_j},$$

this procedure requires on average

$$\mathrm{Compl}_{\mathrm{BS+RMD}}^{\mathrm{rand}}(\mathrm{SDP},\epsilon) = \tilde{\mathcal{O}}\left(\frac{n^2}{\nu^{2.5}} + \frac{mS}{\nu^2}\right)$$

arithmetic operations.

Let ν be defined as above, suppose that $n \geq m$, and fix $\beta \in [2,3]$ such that $n^\beta = mS$. Then, $\mathrm{Compl}_{\mathrm{BS+RMD}}^{\mathrm{rand}}(\mathrm{SDP},\epsilon) \ll \mathrm{Compl}_{\mathrm{IP}}(\mathrm{SDP},\epsilon)$ if

$$\max\left\{1/\nu^{2.5}, n^{\beta-2}/\nu^2\right\} \ll mn^{1.5},$$

where we neglect all logarithmic terms.

Using the alternative problem transformation introduced in Section 8.2, we get to the following complexity estimate. We tag the "Compl"-notion with a "CD" in order to indicate that we perform a Cholesky decomposition of the matrix C in (SDP).

Corollary 8.7 *Let Assumption 8.2 hold. We transform Problem (SDP) into (8.6), apply randomized Mirror-Descent methods to the dual of (8.6), and compute $\hat{X}$ as described in (8.7). On average, we need*

$$\mathrm{Compl}_{\mathrm{CD+RMD}}(\mathrm{SDP},\epsilon\phi^*) = \tilde{\mathcal{O}}\left(\frac{n^2}{\bar{\nu}^{2.5}} + \frac{mn^2}{\bar{\nu}^2} + mn^3\right)$$

arithmetic operations to find a feasible solution to (SDP) with relative accuracy $\epsilon \in (0,1)$, where

$$\bar{\nu} := \frac{\epsilon\lambda_{\min}(C)}{\max_{1\leq k\leq m}\|A_k\|_{(\infty)}\sum_{j=1}^{m}\bar{y}_j}.$$

Note that we write mn^2 instead of mS in the above complexity estimate, as the matrices B_j in (8.6) can be dense, even though the input matrices A_j might have been sparse.

Proof: For any $1 \leq j \leq m$, we write $\lambda_{\max}(B_j)$ as Raleigh quotient and obtain:

$$\lambda_{\max}(B_j) \;=\; \max_{x\neq 0}\frac{x^T B_j x}{x^T x} = \max_{x\neq 0}\frac{\left(L^{-T}x\right)^T A_j\left(L^{-T}x\right)}{x^T x} = \max_{L^{-T}y\neq 0}\frac{y^T A_j y}{y^T C y}$$

$$\begin{aligned} &= \max_{y \neq 0} \frac{y^T A_j y}{y^T C y} \leq \max_{y \neq 0} \frac{y^T A_j y}{y^T y} \max_{y \neq 0} \frac{y^T y}{y^T C y} \\ &= \frac{\lambda_{\max}(A_j)}{\lambda_{\min}(C)} \leq \frac{\|A_j\|_{(\infty)}}{\lambda_{\min}(C)}. \end{aligned}$$

Similarly, we have:

$$-\lambda_{\min}(B_j) = \lambda_{\max}(-B_j) \leq \frac{\|A_j\|_{(\infty)}}{\lambda_{\min}(C)}.$$

We conclude that $\|B_j\|_{(\infty)} \leq \|A_j\|_{(\infty)} / \lambda_{\min}(C)$. It remains to combine Theorem 8.2 with (8.10). $\blacksquare$

Using the same notation as in the above corollary and assuming $n \geq m$, we observe that the relation $\mathrm{Compl}_{\mathrm{CD+RMD}}(\mathrm{SDP}, \epsilon\phi^*) \ll \mathrm{Compl}_{\mathrm{IP}}(\mathrm{SDP}, \epsilon)$ holds if

$$\max\left\{1/\bar{\nu}^{2.5}, m/\bar{\nu}^2\right\} \ll mn^{1.5}.$$

8.4 Interior-Point methods for matrix saddle-point problems

We conclude this chapter by discussing the arithmetic complexity of Interior-Point methods when applied to the new Core Problem (PSP).

Note that we can write Problem (PSP) in the following linear form:

$$\begin{aligned} \min_{X,t} \quad & t \\ \text{s.t.} \quad & -t - \langle D_j, X \rangle_F \leq 0 \qquad \text{for any } j = 1, \ldots, m \\ & \langle \mathbb{I}_n, X \rangle_F = 1 \\ & X \succeq 0. \end{aligned} \qquad (8.11)$$

Applying the same reasoning as in Section 6.2 to the above problem, we observe that Interior-Point (IP) methods require

$$\mathrm{Compl}_{\mathrm{IP}}(\mathrm{PSP}, \epsilon) = \mathcal{O}\left(\sqrt{m+n}\left[mn^3 + m^2 S + m^3\right] \ln\left[(m+n)/\epsilon\right]\right) \quad (8.12)$$

arithmetic operations for finding a feasible ϵ-solution to Problem (PSP).

Comparison: Let us compare finally the complexity results $\mathrm{Compl}_{\mathrm{IP}}(\mathrm{PSP}, \epsilon)$ of Interior-Point methods and $\mathrm{Compl}_{\mathrm{RMD}}^{\mathrm{rand}}(\mathrm{DSP}, \epsilon)$ of randomized Mirror-Descent schemes. In particular, we want to verify that there exists a meaningful class of Problems (PSP), where randomized Mirror-Descent methods outperform Interior-Point methods with respect to the arithmetic complexity. We assume that $n \geq m$. In addition, let $\beta \in [2, 3]$ such that $mS = n^\beta$. Neglecting all logarithmic terms, we have $\mathrm{Compl}_{\mathrm{RMD}}^{\mathrm{rand}}(\mathrm{DSP}, \epsilon) \ll \mathrm{Compl}_{\mathrm{IP}}(\mathrm{PSP}, \epsilon)$ if

$$\mathcal{L}/\epsilon \ll \max\left\{m^{2/5}n^{3/5}, m^{1/2}n^{(3.5-\beta)/2}\right\}. \qquad (8.13)$$

Typically, this condition is satisfied if a large-scale problem needs to be solved up to a very moderate accuracy.

When we focus on the solution accuracy ϵ, we observe that there is a wide gap. Namely, the running of Interior-Point methods is about $\mathcal{O}(\ln[1/\epsilon])$, whereas the complexity estimate of randomized Mirror-Descent algorithms is of the order $\mathcal{O}(1/\epsilon^{2.5})$, where we suppress all parameters except the solution accuracy ϵ in the $\mathcal{O}$-notation. In the next chapters, we will fill this gap by applying advanced First-Order methods such as Smoothing Techniques and Mirror-Prox methods to Problem (DSP).

Chapter 9

Smoothing Techniques for matrix saddle-point problems

$\mathbf{A}$s shown in the last chapter, both (randomized) Mirror-Descent schemes and Interior-Point methods can be applied to Matrix Saddle-Point Problem (PSP), or its Dual (DSP), respectively. We observed that randomized Mirror-Descent schemes are preferably used for solving large-scale problems, as one single iteration of these algorithms can be carried out much faster than in the case of Interior-Point methods. More precisely, randomized Mirror-Descent have – roughly speaking – an iteration cost of $\tilde{\mathcal{O}}\left(n^2[\mathcal{L}/\epsilon]^{0.5} + mS\right)$, while Interior-Point methods need $\mathcal{O}(mn^3 + mS + m^3)$ arithmetic operations per iteration. On the other hand, as the iteration count grows with the order $\mathcal{O}(\mathcal{L}^2 \ln[m]/\epsilon^2)$ for (randomized and deterministic) Mirror-Descent methods, these methods allow the computation of solutions with a relatively low approximation quality $\epsilon > 0$ only. On the contrary, Interior-Point methods can easily derive highly approximate solutions, as the analytical complexity is given by $\mathcal{O}\left([m + n]^{0.5} \ln[(m + n)/\epsilon]\right)$ for these schemes. When we compare the analytical complexity of the two methods, we recognize a dramatic difference with respect to the solution accuracy. In particular, we might hope to find a method that has a more favorable analytical complexity than Mirror-Descent methods, that is, an algorithm whose iteration count grows with of the order $\mathcal{O}(\mathcal{L}/\epsilon)$, but that still outperforms Interior-Points schemes regarding the cost per iteration.

We know from Sections 4.1.1 and 4.1.2 that Smoothing Techniques and Mirror-Prox methods have an analytical complexity that is of the desired order. In this chapter, we apply Smoothing Techniques to Problem (PSP) and its Dual (DSP). As we show in Section 9.1, it is more favorable to apply Smoothing Techniques to (DSP) than to run them on (PSP), because we need only one instead of two costly matrix exponentiations per iteration. In addition, we observe in the same section that the iteration cost of Smoothing Techniques when applied to (DSP) are dominated by the computation of this matrix exponential and some other matrix operations not exceeding the complexity $\mathcal{O}(mS)$, ending up in the requirement of $\mathcal{O}(n^3 + mS)$ arithmetic operations per iteration. As desired, it is indeed less costly to carry out one iteration of Smoothing Techniques than to perform one Newton step in Interior-Point methods. In Section 9.2, we specify the arithmetic complexity of Smoothing Techniques when applied to (DSP). As an immediate consequence, we obtain a complexity estimate for an alternative procedure based on Smoothing Techniques for solving (SDP) approximately. Finally, we verify in the same section that there exists a meaningful range for the problem parameters m, n, and ϵ for which the efficiency estimate of Smoothing Techniques when applied to (DSP) outperforms the arithmetic complexity of both Interior-Point methods and randomized Mirror-Descent algorithms.

Contributions: Nesterov [Nes05] studied the application of Smoothing Techniques to the eigenvalue optimization problem stated in (DSP). Our contribution consists in combining these Smoothing Techniques with the problem transformation described in Section 8.2 in order to get an algorithm for solving (SDP). Similar results were independently and simultaneously obtained by Iyengar et al. [IPS05, IPS11]. They have developed a similar procedure to solve some semidefinite packing problems, applying Nesterov's scheme to a different model than ours. Their construction is based on the Lagrangian relaxation of the initial problem, which, in their first paper [IPS05], originates from semidefinite relaxations of MaxCut and Graph Coloring. In their second paper [IPS11], they generalize their considerations to a broader class of problems.

Relevant literature: We use the reference [Nes05]. Parts of this chapter are taken from [BB09].

9.1 Smooth approximation of the primal and of the dual

We observed in Section 8.3 that the core matrix saddle-point problem of this thesis can take two different forms, namely the Primal Form (PSP) and the Dual Form (DSP). After specifying a smoothing procedure for either problems, we study in this section the different cost-dominating computations per iteration of Algorithm 4.1 when applied to them.

Let $\mu > 0$. We start with the primal problem and endow $\mathbb{R}^m$ with the 1-norm $\|\cdot\|_1$ and choose d_{Δ_m} as distance-generating function on Δ_m. As presented in Section 4.1.1, the objective function $\overline{\varphi}$ of the Primal Problem (PSP) can be approximated by the following smooth function:

$$\overline{\varphi}_{\mu} : \mathcal{S}^n \to \mathbb{R} : X \mapsto \max_{y \in \Delta_m} \left\{ -\sum_{j=1}^{m} y_j \langle D_j, X \rangle_F - \mu d_{\Delta_m}(y) \right\}. \qquad (9.1)$$

We can easily derive a closed-form formula for both the value of this function and its gradient:

$$\overline{\varphi}_{\mu}(X) = \mu \ln \left(\sum_{j=1}^{m} \exp\left(-\langle D_j, X \rangle_F / \mu\right) \right) - \mu \ln(m) \in \mathbb{R}, \qquad X \in \mathcal{S}^n,$$

and:

$$\nabla \overline{\varphi}_{\mu}(X) = -\sum_{j=1}^{m} \frac{D_j \exp\left(-\langle D_j, X \rangle_F / \mu\right)}{\sum_{k=1}^{m} \exp\left(-\langle D_k, X \rangle_F / \mu\right)} \in \mathcal{S}^n, \qquad X \in \mathcal{S}^n;$$

see [Nes05]. The computation of the gradient $\nabla \overline{\varphi}_{\mu}(X)$ requires $\mathcal{O}(mS)$ arithmetic operations.

Now, we equip $\mathcal{S}^n$ with the induced 1-norm $\|\cdot\|_{(1)}$ and fix $d_{\Delta_n^M}$ as distance-generating function on Δ_n^M. We apply Algorithm 4.1 to the problem

$$\min_{X \in \Delta_n^M} \overline{\varphi}_{\mu}(X). \qquad (9.2)$$

At every iteration of this algorithm, we need to find the unique optimal solutions of two(!) optimization problems of the form

$$\min_{X \in \Delta_n^M} \left\{ \langle G, X \rangle_F + d_{\Delta_n^M}(X) \right\},$$

where G is a symmetric matrix. As shown in Example 2.15, finding these optimal solutions corresponds to carrying out two matrix exponentiations, each of them requiring $\mathcal{O}(n^3)$ arithmetic operations. So, the iteration cost of Algorithm 4.1 when applied to (9.2) are dominated by the complexity of the computation of two matrix exponentials and several matrix operations not exceeding the cost $\mathcal{O}(mS)$.

We turn now our attention to the Dual Problem Formulation (DSP). As above, we consider $\mathcal{S}^n$ with the induced 1-norm and choose $d_{\Delta_n^M}$ as distance-generating function on Δ_n^M. The computation cost of the gradient of the smoothed function, namely of:

$$\overline{\phi}_{\mu}(y) := \max_{X \in \Delta_n^M} \left\{ \sum_{j=1}^{m} y_j \langle D_j, X \rangle_F - \mu d_{\Delta_n^M}(X) \right\}, \qquad y \in \mathbb{R}^m,$$

is significantly higher. By Theorem 4.1, we have for any $y \in \mathbb{R}^m$:

$$\left[\nabla \overline{\phi}_\mu(y)\right]_j = \langle D_j, X_*(y) \rangle_F \qquad \forall\, 1 \leq j \leq m,$$

where $X_*(y)$ is the unique maximizer of the above maximization problem. The computation of $X_*(y)$ encompasses the assembly of $\sum_{j=1}^m y_j \langle D_j, X \rangle_F$ and a matrix exponentiation; see Example 2.15. Thus, the evaluation of $\nabla \overline{\phi}_\mu(y)$ comprises one matrix exponentiation and some other matrix operations with a cost that does not exceed $\mathcal{O}(mS)$. In contrast, the resolution of the two optimization subproblems require only $\mathcal{O}(m)$ elementary iterations. Indeed, we equip $\mathbb{R}^n$ with the 1-norm and fix d_{Δ_m} as distance-generating function on Δ_m. The two subproblems are then of the form

$$\min_{y \in \Delta_m} \left\{ \langle g, y \rangle + d_{\Delta_m}(y) \right\}, \qquad g \in \Delta_m,$$

whose minimizers can be computed using $\mathcal{O}(m)$ arithmetic operations; see Example 2.14. This leaves a clear advantage for the dual formulation of the core matrix saddle-point problem if $n^3 \gg mS$, as we save one costly matrix exponentiation per iteration.

9.2 Applying an optimal First-Order method

Let us now evaluate the number of iterations needed by Algorithm 4.1 when applied to the dual problem

$$\min_{y \in \Delta_m} \overline{\phi}_\mu(y), \qquad \overline{\phi}_\mu(y) := \max_{X \in \Delta_n^M} \left\{ \sum_{j=1}^m y_j \langle D_j, X \rangle_F - \mu d_{\Delta_n^M}(X) \right\}. \qquad (9.3)$$

Fix $\mathcal{A} : \mathbb{R}^m \to \mathcal{S}^n : y \mapsto \mathcal{A}(y) := \sum_{j=1}^m y_j D_j$. In view of Theorem 4.2, we are supposed to evaluate the operator norm

$$\begin{aligned}
\|\mathcal{A}\|_{\mathbb{R}^m, \mathcal{S}^n} &:= \max_{y \in \mathbb{R}^m, X \in \mathcal{S}^n} \left\{ \langle \mathcal{A}(y), X \rangle_F : \|u\|_1 = 1, \|X\|_{(1)} = 1 \right\} \\
&= \max_{1 \leq j \leq m} \|D_j\|_{(\infty)} =: \mathcal{L}. \qquad (9.4)
\end{aligned}$$

Applying Theorem 4.1, we can therefore guarantee that the smoothed function $\overline{\phi}_\mu$ has a gradient Lipschitz constant of $L_\mu := \mathcal{L}^2/\mu$. We recall from Example 2.18 that d_{Δ_m} and $d_{\Delta_n^M}$ are bounded on their domains from above by $\ln(m)$ and by $\ln(n)$, respectively. We choose accuracy $\epsilon > 0$ and fix the smoothness parameter μ as follows:

$$\mu := \frac{\epsilon}{2\ln(n)}.$$

Now, we choose

$$T \geq \left\lceil \frac{4\mathcal{L}\sqrt{\ln(m)\ln(n)}}{\epsilon} - 1 \right\rceil$$

and assume that the sequences $(x_t)_{t=0}^{T+1}$, $(u_t)_{t=0}^{T+1}$, $(z_t)_{t=0}^{T}$, $(\hat{x}_t)_{t=0}^{T+1}$, and $(L_t)_{t=0}^{T}$ are defined by Algorithm 4.1 when applied to Problem (9.3) with the above smoothness parameter. In addition, we recall that we write $X_*(y)$ for the optimal solution to the maximization problem that describes $\overline{\phi}_\mu(y)$, where $y \in \mathbb{R}^m$. Defining

$$\bar{y} := u_T \in \Delta_m \qquad \text{and} \qquad \bar{X} := \sum_{t=0}^{T} \frac{2(t+1)}{(T+1)(T+2)} X_*(x_t) \in \Delta_n^M,$$

we obtain by Theorem 4.2:

$$0 \leq \overline{\phi}(\bar{y}) - \min_{1 \leq j \leq m} \left\langle D_j, \bar{X} \right\rangle_F \leq \epsilon - \frac{4\chi_T}{(T+1)^2}$$

with

$$\chi_T := \sum_{t=0}^{T-1} (L_{t+1} - L_t) \left(d_{\Delta_m}(z_{t+1}) - \|z_t - \hat{x}_{t+1}\|_1^2 / 2 \right).$$

We note that χ_T vanishes if $L_t = \mathcal{L}$ for any $0 \leq t \leq T + 1$. We conclude:

Theorem 9.1 (Consequence of Section 4 in [Nes05]) *Smoothing Techniques* (ST) *require at most*

$$\mathrm{Compl}_{\mathrm{ST}}(\mathrm{DSP}, \epsilon) = \mathcal{O}\left(\frac{\left[n^3 + mS\right] \mathcal{L}\sqrt{\ln[m]\ln[n]}}{\epsilon} \right)$$

arithmetic operations to compute a pair $(\bar{y}, \bar{X}) \in \Delta_m \times \Delta_n^M$ *for which the difference* $\overline{\phi}(\bar{y}) - \min_{1 \leq j \leq m} \left\langle D_j, \bar{X} \right\rangle_F$ *is bounded from above by* $\epsilon > 0$.

Comparison: We suppose $m \leq n$, fix $\beta \in [2, 3]$ such that $mS = n^\beta$, and neglect all logarithmic terms. We have $\mathrm{Compl}_{\mathrm{ST}}(\mathrm{DSP}, \epsilon) \ll \mathrm{Compl}_{\mathrm{IP}}(\mathrm{PSP}, \epsilon)$ if

$$\mathcal{L}/\epsilon \ll mn^{1/2}, \tag{9.5}$$

and $\mathrm{Compl}_{\mathrm{ST}}(\mathrm{DSP}, \epsilon) \ll \mathrm{Compl}_{\mathrm{RMD}}^{\mathrm{rand}}(\mathrm{DSP}, \epsilon)$ if

$$\min\left\{ n^{3/4}, n^{3-\beta} \right\} \ll \mathcal{L}/\epsilon.$$

That is, $\text{Compl}_{\text{ST}}(\text{DSP}, \epsilon) \ll \min\left\{\text{Compl}_{\text{IP}}(\text{PSP}, \epsilon), \text{Compl}_{\text{RMD}}^{\text{rand}}(\text{DSP}, \epsilon)\right\}$ if

$$\min\left\{n^{3/4}, n^{3-\beta}\right\} \ll \mathcal{L}/\epsilon \ll mn^{1/2}. \tag{9.6}$$

Note that Condition (9.6) cannot be satisfied if $\beta < 2.5$, that is, if – for instance – the matrices D_j are sparse. As Theorem 9.1 shows, sparsity of the matrices D_j does not affect the arithmetic complexity of Smoothing Techniques, as $mS\mathcal{L}/\epsilon = n^{\beta}\mathcal{L}/\epsilon$ is dominated by $n^3\mathcal{L}/\epsilon$, that is, the cost per iteration is dominated by the efficiency of the matrix exponentiation. This is in sharp contrast to the situation of randomized Mirror-Descent methods, where we avoid this costly computation. In particular, the quantity $mS\mathcal{L}^2/\epsilon^2 = n^{\beta}\mathcal{L}^2/\epsilon^2$ in Theorem 8.2 cannot be bounded from above by $n^2\mathcal{L}^{2.5}/\epsilon^{2.5}$. Clearly, in order to enhance the arithmetic complexity of advanced First-Order methods such as Smoothing Techniques when applied to (DSP), we thus need to find a strategy to circumvent the cost-determinating matrix exponentiation, which we standardly do through an eigendecomposition of the argument. The next chapter addresses this topic.

Arithmetic complexity for (SDP): We conclude with the following consequences of Theorem 9.1.

Corollary 9.1 *Let ν be defined as in Corollary 8.6. Applying the same procedure as in Corollary 8.6, but using Smoothing Techniques instead of randomized Mirror-Descent schemes, we need*

$$\text{Compl}_{\text{BS+ST}}(\text{SDP}, \epsilon) = \mathcal{O}\left(\frac{\left[n^3 + mS\right]\sqrt{\ln[m]\ln[n]}}{\nu} \ln\left[\frac{\sum_{j=1}^m \bar{y}_j}{\epsilon}\right]\right)$$

arithmetic operations to find a feasible ϵ-solution to (SDP).

Corollary 9.2 *Assume $\bar{\nu}$ to be given as in Corollary 8.7. Performing the same steps as in Corollary 8.7, but with Smoothing Techniques in the role of randomized Mirror-Descent methods, we are supposed to carry out about*

$$\text{Compl}_{\text{CD+ST}}(\text{SDP}, \epsilon\phi^*) = \mathcal{O}\left(\frac{\left[n^3 + mn^2\right]\sqrt{\ln[m]\ln[n]}}{\bar{\nu}} + mn^3\right)$$

arithmetic operations for deriving a feasible solution to (SDP) with a relative accuracy $\epsilon \in (0, 1)$.

Chapter 10

Applying randomized Mirror-Prox methods

$\mathbf{W}$HEN we compare randomized Mirror-Descent schemes with advanced First-Order methods such as Smoothing Techniques, we make two important observations. On the hand, the analytical complexity of randomized Mirror-Descent schemes grows – with respect to the solution accuracy – one order magnitude faster than advanced First-Order methods. On the other hand, randomized Mirror-Descent methods have the significant advantage of avoiding the need of costly eigendecompositions when applied to (DSP). This is in sharp contrast to standard Smoothing Techniques, where we need to compute a matrix exponential through an eigendecomposition at every iteration. In this chapter, we introduce a procedure that combines the advantages of the two methods.

As we noticed in Sections 4.1.1 and 4.1.2, original Mirror-Prox methods (see Algorithm 4.2) have the same analytical complexity as Smoothing Techniques. We specify the setting of Mirror-Prox methods when applied to (DSP) in Section 10.1, and verify in Section 10.2 that they have the same arithmetic complexity as Smoothing Techniques for (DSP).

In particular, Smoothing Techniques and Mirror-Prox methods share the same computational bottleneck. At every iteration of Mirror-Prox methods, we are supposed to carry out two matrix exponentiations, which dominate the iteration cost provided that $mS \ll n^3$. In Section 10.3, we introduce a randomization procedure for computing matrix exponentials. This procedure is based on a vector sampling and on an appropriate truncation of the Taylor series approximation of the true matrix exponential. The computation of this random approximation can be carried out by performing matrix-vector products only. As we show in Section 10.3, we obtain a method that outperforms all its competitors with the respect to the arithmetic complexity on a reasonably large subfamily of (DSP). However, as we observe at the end of this chapter, the arithmetic complexity result of randomized Mirror-Prox methods cannot be transformed into an efficiency estimate for the initial Problem (SDP).

Note that Arora and Kale [AK07] introduced a randomization procedure for the computation of matrix exponentials in Algorithm 7.3. They consider also an appropriate truncation of the exponential Taylor series approximation, but project it on, roughly speaking, $\mathcal{O}(1/\epsilon^2)$ random directions using Johnson-Lindenstrauss Lemma [JL84], where $\epsilon > 0$ denotes the accuracy. With the randomization procedure that is described in this chapter, we need to project the truncated Taylor series approximation only on, again roughly speaking, $\mathcal{O}(1/\epsilon)$ random directions.

Contributions and relevant literature: The material of this chapter is taken from [BBN11]. Both the randomization procedure for Mirror-Prox methods (described in Section 10.3.1) and the complexity analysis of the resulting algorithm were originally introduced in this joint paper. We are indebted to Arkadi Nemirovski for establishing the most relevant theoretical results in this project, that is, for the randomization procedure shown in Section 10.3.1, Proposition 10.1, and most of the results presented in Section 10.3.2. In this joint project, we were concerned mainly with the numerical testing of the method. The numerical results will be shown in the next chapter.

For Section 10.2, we also need the reference [Nem04a].

10.1 Setup of the algorithm

As for randomized Mirror-Descent schemes in Section 8.3.2 and for Smoothing Techniques in Chapter 9, we consider the following matrix saddle-point problem:

$$\phi^* = \min_{y \in \Delta_m} \max_{X \in \Delta_n^M} \phi(y, X), \qquad \phi(y, X) := \sum_{j=1}^{m} y_j \langle D_j, X \rangle_F, \qquad \text{(DSP)}$$

where $D_1, \ldots, D_n$ are symmetric $(n \times n)$-matrices, each of them with at most S non-zero entries. We suppose that $mS \geq n^2$, that is, we allow the matrices $\sum_{j=1}^{m} y_j D_j$ to be fully dense. We recall that we are interested in large-scale problems, implying that we clearly have $n, m \geq 3$. We make these lower bounds explicit, as they are needed in the proof of Proposition 10.1 and in Section 10.3.2. In addition, note that we write Problem (DSP) in a slightly different form than in the two previous chapters in order to reveal its saddle-point structure more clearly.

The function $\phi(y, X)$ can be written as $\phi(y, X) = \langle \mathcal{A}(y), X \rangle_F$ with the linear operator $\mathcal{A} : \mathbb{R}^m \to \mathcal{S}^n : y \mapsto \sum_{j=1}^m D_j y_j$. As explained in Section 4.1.2, Mirror-Prox methods deal with the following linear operator:

$$F(y, X) := \left(\frac{\partial \phi(y, X)}{\partial y}, -\frac{\partial \phi(y, X)}{\partial X} \right) = (\mathcal{A}^*(X), -\mathcal{A}(y))$$

for any $(y, X) \in \Delta_m \times \Delta_n^M$, where the linear mapping

$$\mathcal{A}^* : \mathcal{S}^n \to \mathbb{R}^m : X \mapsto \mathcal{A}^*(X) = (\langle D_j, X \rangle_F)_{j=1}^m$$

is the conjugate of $\mathcal{A}$.

As usual for the (matrix) simplex setup, we equip $\mathbb{R}^m$ and $\mathcal{S}^n$ with the norms $\|\cdot\|_1$ and $\|\cdot\|_{(1)}$, respectively. In addition, we choose d_{Δ_m} and $d_{\Delta_n^M}$ as distance-generating functions on the sets Δ_m and Δ_n^M, respectively. We recall from (9.4) that the norm of the operator $\mathcal{A}$ takes the value

$$\|\mathcal{A}\|_{\mathbb{R}^m, \mathcal{S}^n} = \max_{1 \leq j \leq m} \|D_j\|_{(\infty)} =: \mathcal{L}$$

for this setup.

Mirror-Prox methods consider the product set $\Delta_m \times \Delta_n^M$, which is a subset of the space $\mathbb{R}^m \times \mathcal{S}^n$. In view of Section 4.1.2, we equip this space with the norm

$$\|(y, X)\|_{\mathbb{R}^m \times \mathcal{S}^n} := \sqrt{\frac{1}{2\ln(m)} \|y\|_1^2 + \frac{1}{2\ln(n)} \|X\|_{(1)}^2}$$

for any $(y, X) \in \mathbb{R}^m \times \mathcal{S}^n$, for which the dual norm is given by

$$\|(u, V)\|_{\mathbb{R}^m \times \mathcal{S}^n, *} := \sqrt{2\ln(m) \|u\|_\infty^2 + 2\ln(n) \|V\|_{(\infty)}^2} \tag{10.1}$$

for each $(u, V) \in \mathbb{R}^m \times \mathcal{S}^n$. Applying Lemma 4.1, we observe that the linear operator F is Lipschitz continuous with the Lipschitz constant

$$L := 2\mathcal{L}\sqrt{\ln(m)\ln(n)}.$$

Finally, as presented in Section 4.1.2, the mapping

$$d_{\Delta_m \times \Delta_n^M} : \Delta_m \times \Delta_n^M \to \mathbb{R} : (y, X) \mapsto \frac{1}{2\ln(m)} d_{\Delta_m}(y) + \frac{1}{2\ln(n)} d_{\Delta_n^M}(X)$$

represents a distance-generating function on $\Delta_m \times \Delta_n^M$. We have for this distance-generating function:

$$\Omega_V(d_{\Delta_m \times \Delta_n^M}) = \sqrt{2} \qquad \text{and} \qquad c(d_{\Delta_m \times \Delta_n^M}) = (c(d_{\Delta_m}), c(d_{\Delta_n^M}));$$

see Equation (4.9) and Example 2.16 for the definition of the d-centers $c(d_{\Delta_m})$ and $c(d_{\Delta_n^M})$.

10.2 Complexity of deterministic Mirror-Prox methods

We study in this section the arithmetic complexity of original Mirror-Prox methods when applied to Problem (DSP). Not surprisingly, we observe that this complexity result coincides with the efficiency estimate for Smoothing Techniques.

We use the original Mirror-Prox methods described in Algorithm 4.2 for solving approximately Problem (DSP). With the setting that we introduced in the previous section, we end up with a method that has the form of Algorithm 10.1. In view of Theorem 4.3, we need to perform

$$T \geq \frac{2\mathcal{L}\sqrt{\ln(m)\ln(n)}}{\epsilon}$$

iterations of Algorithm 10.1 in order to find a pair $(\bar{y}, \bar{X}) \in \Delta_m \times \Delta_n^M$ for which

$$\max_{X \in \Delta_n^M} \phi(\bar{y}, X) - \min_{y \in \Delta_m} \phi(y, \bar{X}) \leq \epsilon, \tag{10.2}$$

where $\epsilon > 0$. The cost of any iteration of this method is determined by the following computations:

1. Given two matrices $X, \bar{X} \in \Delta_n^M$, we need to evaluate $\mathcal{A}^*(X)$ and $\mathcal{A}^*(\bar{X})$, requiring $\mathcal{O}(mS)$ arithmetic operations.

2. We need another $\mathcal{O}(mS)$ arithmetic operations for the computation of $-\mathcal{A}(y)$ and $-\mathcal{A}(\bar{y})$, where $y, \bar{y} \in \Delta_m$ are given.

3. We are supposed to carry out two matrix exponentiations, coming with the cost of $\mathcal{O}(n^3)$ arithmetic operations when performing these operations through eigendecompositions of the matrices at hand.

Note that the computation of y_t and $\bar{y}_t$ in Algorithm 10.1 is complexity-wise unproblematic, as each of them requires only $\mathcal{O}(m)$ arithmetic operations; see Example 2.14. We have verified the following complexity result.

Theorem 10.1 (Section 5 in [Nem04a]) *Choose* $\epsilon > 0$. *Mirror-Prox* (MP) *methods require*

$$\mathrm{Compl}_{\mathrm{MP}}(\mathrm{DSP}, \epsilon) = \mathcal{O}\left(\frac{\left[n^3 + mS\right]\mathcal{L}\sqrt{\ln(m)\ln(n)}}{\epsilon}\right)$$

arithmetic operations for the computation of a pair $(\bar{y}, \bar{X}) \in \Delta_m \times \Delta_n^M$ *that satisfies (10.2).*

Algorithm 10.1 Mirror-Prox methods applied to (DSP)

1: Choose $T \in \mathbb{N}$.
2: Set $y_0 := c(d_{\Delta_m})$ and let $V_0 \in \mathcal{S}_n$ be the all zero matrix.
3: **for** $1 \leq t \leq T$ **do**
4:    Choose $\gamma_t > 0$ such that $\gamma_t \leq \left(2\mathcal{L}\sqrt{\ln(m)\ln(n)}\right)^{-1}$.
5:    Compute

$$F\left(y_{t-1}, \frac{\exp(V_{t-1})}{\mathrm{Tr}(\exp(V_{t-1}))}\right) = \left(\frac{\mathcal{A}^*(\exp(V_{t-1}))}{\mathrm{Tr}(\exp(V_{t-1}))}, -\mathcal{A}(y_{t-1})\right)$$
$$=: \ (g(V_{t-1}), H(y_{t-1})).$$

6:    Set

$$\bar{y}_t = \arg\min_{y \in \Delta_m} \left\{\langle 2\ln(m)\gamma_t g(V_{t-1}) - d'_{\Delta_m}(y_{t-1}), y\rangle + d_{\Delta_m}(y)\right\}.$$

7:    Define $\bar{V}_t = V_{t-1} - 2\gamma_t \ln(n) H(y_{t-1})$.
8:    Compute

$$F\left(\bar{y}_t, \frac{\exp(\bar{V}_t)}{\mathrm{Tr}(\exp(\bar{V}_t))}\right) = \left(\frac{\mathcal{A}^*(\exp(\bar{V}_t))}{\mathrm{Tr}(\exp(\bar{V}_t))}, -\mathcal{A}(\bar{y}_t)\right) =: \left(g(\bar{V}_t), H(\bar{y}_t)\right).$$

9:    Set $y_t = \arg\min_{y \in \Delta_m} \left\{\langle 2\ln(m)\gamma_t g(\bar{V}_t) - d'_{\Delta_m}(y_{t-1}), y\rangle + d_{\Delta_m}(y)\right\}$.
10:   Define $V_t = V_{t-1} - 2\gamma_t \ln(n) H(\bar{y}_t)$.
11: **end for**
12: Return $(\bar{y}, \bar{X}) := \left(\sum_{t=1}^{T} \gamma_t\right)^{-1} \sum_{t=1}^{T} \gamma_t \left(\bar{y}_t, \exp(\bar{V}_t)/\mathrm{Tr}[\exp(\bar{V}_t)]\right)$.

10.3 Randomized Mirror-Prox methods

In the last section, we observed that the cost per iteration of Mirror-Prox methods when applied to (DSP) is dominated by the computation of two matrix exponentials, provided that $mS \ll n^3$. In the remainder of this chapter, we derive now a Mirror-Prox algorithm that uses random estimates instead of the precise values of these matrix exponentials. Importantly, these random approximations can be computed by performing matrix-vector products only, resulting in a method that requires for many problems on average less arithmetic operations than its original counterpart.

In view of Algorithm 10.1, these cost-critical matrix exponentials are only needed to compute the first component of linear operator F. We intend to replace this first component of the operator F, that is,

$$g(V) := \frac{\mathcal{A}^*(\exp(V))}{\mathrm{Tr}(\exp(V))}, \qquad V \in \mathcal{S}^n,$$

by a random estimate $\hat{g}_\xi(V)$. We use the precise value of the second component of F, as its computation is unproblematic.

10.3.1 Randomization strategy

Let us describe the randomization procedure we use. We choose $V \in \mathcal{S}^n$ and observe that we can write $\exp(V)$ as the product $\exp(V/2)\exp(V/2)$. For the time being, let us assume that we have access to the exact value of $\exp(V/2)$. We denote by $(\Xi_\zeta, \mathcal{B}_\zeta, \mathbb{P}_\zeta)$ a Borel probability space and assume that the random vectors $\zeta_i : \Xi_\zeta \to \mathbb{R}^n$ are independent and $\mathcal{N}(0, \mathbb{I}_n)$-distributed for any $1 \leq i \leq N$. In addition, we denote by $(\Xi, \mathcal{B}, \mathbb{P}) := (\otimes_{i=1}^N \Xi_\zeta, \otimes_{i=1}^N \mathcal{B}_\zeta, \otimes_{i=1}^N \mathbb{P}_\zeta)$ the corresponding product space and write ξ for the sequence

$$(\zeta_1, \ldots, \zeta_N) : \Xi \to \mathbb{R}^{n \times N}$$

of random vectors.

Defining $\chi_i := \exp(V/2)\zeta_i$ for any $1 \leq i \leq N$, we observe that:

$$\mathbb{E}_\mathbb{P}\left[\frac{1}{N} \sum_{i=1}^N \left[\chi_i^T D_1 \chi_i; \ldots; \chi_i^T D_m \chi_i \right] \right] = \mathcal{A}^*(\exp(V)) = \left(\langle D_j, \exp(V) \rangle_F \right)_{j=1}^m,$$

and that:

$$\mathbb{E}_\mathbb{P}\left[\frac{1}{N} \sum_{i=1}^N \chi_i^T \chi_i \right] = \mathrm{Tr}(\exp(V));$$

see Part a) of Lemma B.3 for the proofs. That is, the random vectors χ_i can be used to define unbiased estimators of both the numerator and the denumerator of $\mathcal{A}^*(\exp(V))/\mathrm{Tr}(\exp(V))$. Motivated by these observations, we use

$$g_\xi(V) := \frac{\sum_{i=1}^N \left[\chi_i^T D_1 \chi_i; \ldots; \chi_i^T D_m \chi_i \right]}{\sum_{i=1}^N \chi_i^T \chi_i} \tag{10.3}$$

as a random estimate of $g(V) = \mathcal{A}^*(\exp(V))/\mathrm{Tr}(\exp(V))$. For this random estimate, we have the following guarantees.

Proposition 10.1 *For any $V \in \mathcal{S}^n$, we have:*

$$\| \mathbb{E}_\mathbb{P}[g_\xi(V)] - g(V) \|_\infty \leq \mathcal{O}(1)\mathcal{L}\sqrt{\ln(m)}/N \tag{10.4}$$

and:

$$\mathbb{E}_\mathbb{P}\left[\| g_\xi(V) - \mathbb{E}_\mathbb{P}[g_\xi(V)] \|_\infty^2 \right] \leq \mathcal{O}(1)\mathcal{L}^2 \ln(m)/N. \tag{10.5}$$

The proof of Proposition 10.1 is highly technical and is based on mathematical concepts that are completely different than the ones we have used so far in this thesis. We give its proof in Appendix B.3.

The above estimate $g_\xi(V)$ requires the random vectors $\chi_i = \exp(V/2)\zeta_i$ for $1 \le i \le N$. Given the matrix V and the random vectors ζ_i, it remains to discuss a strategy to compute – or to approximate – the random vectors χ_i by carrying out matrix-vector products only. Instead of using the exact random vectors χ_i, we suggest to replace them by the random vectors

$$\bar\chi_i := \sum_{k=0}^{K} \frac{1}{k!} (V/2)^k \zeta_i \qquad \forall\, 1 \le i \le N, \tag{10.6}$$

with $K \in \mathbb{N}$ large enough to guarantee a high approximation quality. The approximation quality of the **truncated Taylor series** is given by the following proposition.

Proposition 10.2 *Let $W \in \mathcal{S}^n$ and $K \ge \exp(2)\,\|W\|_{(\infty)}$. Then,*

$$\left\| \exp(W) - \sum_{k=0}^{K} \frac{W^k}{k!} \right\|_{(\infty)} \le \exp(-K).$$

This result can be proven by applying the same arguments as in the proof of Lemma 6 in [AK07].

Note that we can compute $\bar\chi_i$ in a recursive way. Namely, we set $\bar v_0 := \zeta_i$, and define $\bar v_k := \frac{1}{2k} V \bar v_{k-1}$ for any $1 \le k \le K$. The computation of $\bar\chi_i$ can thus be carried out by performing matrix-vector products solely, requiring in total $\mathcal{O}(Kn^2)$ arithmetic operations. If the matrix V is sparse, say it has $\bar S$ non-zero entries, this complexity estimate reduces to $\mathcal{O}(K\bar S)$.

We end up with

$$\hat g_{\xi,K}(V) := \frac{\sum_{i=1}^{N} \left[\bar\chi_i^T D_1 \bar\chi_i; \ldots; \bar\chi_i^T D_m \bar\chi_i \right]}{\sum_{i=1}^{N} \bar\chi_i^T \bar\chi_i} \tag{10.7}$$

as random estimate for $g(V)$. The computation of this approximation requires in total $\mathcal{O}(KNn^2 + mS)$ arithmetic operations. In addition, it can be written as

$$\hat g_{\xi,K}(V) = \frac{\mathcal{A}^*(\hat{\mathcal{E}}_{\xi,K}(V))}{\mathrm{Tr}(\hat{\mathcal{E}}_{\xi,K}(V))}, \qquad \hat{\mathcal{E}}_{\xi,K}(V) := \frac{1}{N} \sum_{i=1}^{N} \bar\chi_i \bar\chi_i^T. \tag{10.8}$$

Algorithm 10.2 Randomized Mirror-Prox methods for (DSP)

1: Choose the number of iterations $T \in \mathbb{N}$, the sample size $N \in \mathbb{N}$, and truncation levels $K_t \in \mathbb{N}$ for any $1 \leq t \leq T$.

2: Generate $2T$ independent samples $\xi_1, \ldots, \xi_{2T}$, each of them comprising N independent realizations $\zeta_{t,1}, \ldots, \zeta_{t,N}$ of a $\mathcal{N}(0, \mathbb{I}_n)$-distributed random vector.

3: Set $y_0 := c(d_{\Delta_m})$ and let $V_0 \in \mathcal{S}_n$ be the all zero matrix.

4: **for** $1 \leq t \leq T$ **do**

5:     Choose $\gamma_t > 0$ such that

$$\gamma_t \leq \left(2\mathcal{L}\sqrt{2\ln(m)\ln(n)} \right)^{-1}. \tag{10.9}$$

6:     Approximate $F\left(y_{t-1}, \exp(V_{t-1})/\mathrm{Tr}[\exp(V_{t-1})]\right)$ by $\left(\hat{g}_{\xi_{2t-1}, K_t}(V_{t-1}), H(y_{t-1})\right)$, where $\hat{g}_{\xi_{2t-1}, K_t}(V_{t-1})$ is given by (10.7) and $H(y_{t-1}) := -\mathcal{A}(y_{t-1})$.

7:     Compute the minimizer of

$$y \mapsto \left\langle 2\ln(m)\gamma_t \hat{g}_{\xi_{2t-1}, K_t}(V_{t-1}) - d'_{\Delta_m}(y_{t-1}), y \right\rangle + d_{\Delta_m}(y)$$

over Δ_m and denote it by $\bar{y}_t$.

8:     Define $\bar{V}_t = V_{t-1} - 2\gamma_t \ln(n) H(y_{t-1})$.

9:     Approximate $F\left(\bar{y}_t, \exp(\bar{V}_t)/\mathrm{Tr}[\exp(\bar{V}_t)]\right)$ by $\left(\hat{g}_{\xi_{2t}, K_t}(\bar{V}_t), H(\bar{y}_t)\right)$, where $\hat{g}_{\xi_{2t}, K_t}(\bar{V}_t)$ is given by (10.7) and $H(\bar{y}_t) := -\mathcal{A}(\bar{y}_t)$.

10:    Set

$$y_t = \arg \min_{y \in \Delta_m} \left\{ \left\langle 2\ln(m)\gamma_t \hat{g}_{\xi_{2t}, K_t}(\bar{V}_t) - d'_{\Delta_m}(y_{t-1}), y \right\rangle + d_{\Delta_m}(y) \right\}.$$

11:    Define $V_t = V_{t-1} - 2\gamma_t \ln(n) H(\bar{y}_t)$.

12: **end for**

13: Return $\bar{y} := \left(\sum_{t=1}^{T} \gamma_t \right)^{-1} \sum_{t=1}^{T} \gamma_t \bar{y}_t.$

Using the outlined randomization procedure and running Mirror-Prox methods with noisy first-order information (Algorithm 4.3) on (DSP) with the setting described in Section 10.1, we obtain a method as described in Algorithm 10.2.

10.3.2 Complexity of randomized Mirror-Prox schemes

Let us now derive an arithmetic complexity result for Algorithm 10.2. We use the notation of Algorithm 10.2.

10.3.2.1 Discussion of the truncation levels

Choose the number of iterations $T \in \mathbb{N}$ and let $1 \leq t \leq T$. We start by fixing the truncation levels $K_t \in \mathbb{N}$ of the Taylor approximations of the matrix exponentials $\exp(V_t/2)$ and $\exp(\bar{V}_t/2)$. Using the recursive definitions of V_t and $\bar{V}_t$ in Algorithm 10.2 and considering that V_0 is the all zero matrix, we have:

$$\|V_t\|_{(\infty)} \leq 2\ln(n)\mathcal{L}\sum_{\tau=1}^{t}\gamma_\tau \leq t\sqrt{\frac{\ln(n)}{2\ln(m)}} \quad \text{and} \quad \|\bar{V}_t\|_{(\infty)} \leq t\sqrt{\frac{\ln(n)}{2\ln(m)}}.$$

Let $0 < \epsilon \ll 1$ and $K_t = \mathcal{O}(1)t\sqrt{\ln(n)/\ln(m)}\ln(1/\epsilon)$, where $\mathcal{O}(1)$ denotes an appropriate absolute constant. According to Proposition 10.2, the norm of the difference between the exact matrix exponentials $\exp(V_t/2)$ and $\exp(\bar{V}_t/2)$ and their truncated Taylor series approximations is at most ϵ. Let us now initialize ϵ as machine accuracy. Setting

$$K_t = \mathcal{O}(1)t\sqrt{\ln(n)/\ln(m)} \tag{10.10}$$

with a moderate constant[1] $\mathcal{O}(1)$, we observe that the truncated Taylor series approximation is – for all practical applications – the same as the true matrix exponential. Justified by this observation, we will perform a simplified analysis of Algorithm 10.2, where we do not elaborate on the difference between g_ξ defined in (10.3) and on its approximation $\hat{g}_{\xi,K}$ given in (10.7). That is, we study the arithmetic complexity of an idealized version of Algorithm 10.2, where we use g_ξ instead of $\hat{g}_{\xi,K}$.

10.3.2.2 Expected convergence

The expected convergence of the idealized version of Algorithm 10.2 can be deduced from Theorem 4.4. In this section, we bound from above all quantities that define ϵ in Theorem 4.4.

Let $1 \leq t \leq T$ and recall that we write $\xi_{[t]}$ for the sequence $(\xi_1, \ldots, \xi_t)$. Using the notation of Algorithms 4.3 and 10.2, we have for the Definitions (4.14):

$$
\begin{aligned}
z_{t-1} &= \left(y_{t-1}, \exp(V_{t-1})/\mathrm{Tr}[\exp(V_{t-1})]\right) \\
w_t &= \left(\bar{y}_t, \exp(\bar{V}_t)/\mathrm{Tr}[\exp(\bar{V}_t)]\right) \\
\mu_{z_{t-1}} &= \left(\mathbb{E}_\mathbb{P}\left[g_{\xi_{2t-1}}(V_{t-1})|\xi_{[2t-2]}\right] - g(V_{t-1}), 0\right) =: (\bar{\mu}_{z_{t-1}}, 0) \\
\mu_{w_t} &= \left(\mathbb{E}_\mathbb{P}\left[g_{\xi_{2t}}(\bar{V}_t)|\xi_{[2t-1]}\right] - g(\bar{V}_t), 0\right) =: (\bar{\mu}_{w_t}, 0) \\
\sigma_{z_{t-1}} &= \left(g_{\xi_{2t-1}}(V_{t-1}) - \mathbb{E}_\mathbb{P}\left[g_{\xi_{2t-1}}(V_{t-1})|\xi_{[2t-2]}\right], 0\right) =: (\bar{\sigma}_{z_{t-1}}, 0) \\
\sigma_{w_t} &= \left(g_{\xi_{2t}}(\bar{V}_t) - \mathbb{E}_\mathbb{P}\left[g_{\xi_{2t}}(\bar{V}_t)|\xi_{[2t-1]}\right], 0\right) =: (\bar{\sigma}_{w_t}, 0),
\end{aligned}
$$

---

[1] With the standard machine accuracy of Matlab, which is 2^{-52}, this moderate constant is given by $\ln(2^{52})\exp(2)/(2\sqrt{2}) < 95$.

where we set $\mathbb{E}_\mathbb{P}\left[g_{\xi_1}(V_0)|\xi_{[0]}\right] = \mathbb{E}_\mathbb{P}\left[g_{\xi_1}(V_0)\right]$. Exploiting the definition of the dual norm (see (10.1)) and Inequality (10.4), we obtain:

$$\begin{aligned} \mathbb{E}_\mathbb{P}\left[\left\|\mu_{z_{t-1}}\right\|_{\mathbb{R}^m\times\mathcal{S}^n,*}|\xi_{[2t-2]}\right] &= \sqrt{2\ln(m)}\mathbb{E}_\mathbb{P}\left[\left\|\bar\mu_{z_{t-1}}\right\|_\infty|\xi_{[2t-2]}\right] \\ &\leq \mathcal{O}(1)\mathcal{L}\ln(m)/N. \end{aligned}$$

In a similar way, we can derive the following inequality:

$$\begin{aligned} \mathbb{E}_\mathbb{P}\left[\left\|\mu_{w_t}\right\|_{\mathbb{R}^m\times\mathcal{S}^n,*}|\xi_{[2t-1]}\right] &= \sqrt{2\ln(m)}\mathbb{E}_\mathbb{P}\left[\left\|\bar\mu_{w_t}\right\|_\infty|\xi_{[2t-1]}\right] \\ &\leq \mathcal{O}(1)\mathcal{L}\ln(m)/N. \end{aligned} \tag{10.11}$$

Recall that γ_t denotes the step-size at iteration t of Algorithm 10.2. Using the above bounds, we obtain:

$$\mathbb{E}_{\mathbb{P}^{2T}}\left[\sum_{t=1}^T \gamma_t^2 \left\|\mu_{w_t} - \mu_{z_{t-1}}\right\|^2_{\mathbb{R}^m\times\mathcal{S}^n,*}\right] \leq \mathcal{O}(1)\mathcal{L}^2\ln^2(m)\sum_{t=1}^T \gamma_t^2/N^2. \tag{10.12}$$

Finally, Inequality (10.11) leads to

$$\mathbb{E}_{\mathbb{P}^{2T}}\left[\sum_{t=1}^T \gamma_t \left\|\mu_{w_t}\right\|_{\mathbb{R}^m\times\mathcal{S}^n,*}\right] \leq \mathcal{O}(1)\mathcal{L}\ln(m)\sum_{t=1}^T \gamma_t/N. \tag{10.13}$$

For $\bar\sigma_{z_{t-1}}$ and $\bar\sigma_{w_t}$, we have by Inequality (10.5):

$$\mathbb{E}_\mathbb{P}\left[\left\|\bar\sigma_{z_{t-1}}\right\|^2_\infty|\xi_{[2t-2]}\right] \leq \mathcal{O}(1)\mathcal{L}^2\ln(m)/N,$$

and:

$$\mathbb{E}_\mathbb{P}\left[\left\|\bar\sigma_{w_t}\right\|^2_\infty|\xi_{[2t-1]}\right] \leq \mathcal{O}(1)\mathcal{L}^2\ln(m)/N. \tag{10.14}$$

As an immediate consequence, we obtain:

$$\begin{aligned} \mathbb{E}_{\mathbb{P}^{2T}}&\left[\sum_{t=1}^T \gamma_t^2 \left\|\sigma_{w_t} - \sigma_{z_{t-1}}\right\|^2_{\mathbb{R}^m\times\mathcal{S}^n,*}\right] \\ &= 2\ln(m)\mathbb{E}_{\mathbb{P}^{2T}}\left[\sum_{t=1}^T \gamma_t^2 \left\|\bar\sigma_{w_t} - \bar\sigma_{z_{t-1}}\right\|^2_\infty\right] \\ &\leq \mathcal{O}(1)\mathcal{L}^2\ln^2(m)\sum_{t=1}^T \gamma_t^2/N. \end{aligned} \tag{10.15}$$

Observe that $\mathbb{E}_\mathbb{P}\left[\bar\sigma_{z_{t-1}}|\xi_{[2t-2]}\right] = \mathbb{E}_\mathbb{P}\left[\bar\sigma_{w_t}|\xi_{[2t-1]}\right] = 0$. Considering Example A.2, applying Proposition A.1, and using (10.14), it holds that:

$$\mathbb{E}_{\mathbb{P}^{2T}}\left[\left\|\sum_{t=1}^T \gamma_t\bar\sigma_{w_t}\right\|^2_\infty\right] \leq \mathcal{O}(1)\mathcal{L}^2\ln^2(m)\sum_{t=1}^T \gamma_t^2/N.$$

Due to Cauchy-Schwarz Inequality, we end up with the relation:

$$\mathbb{E}_{\mathbb{P}2T}\left[\left\|\sum_{t=1}^{T}\gamma_t\sigma_{w_t}\right\|_{\mathbb{R}^m\times\mathcal{S}^n,*}\right] = \sqrt{2\ln(m)}\,\mathbb{E}_{\mathbb{P}2T}\left[\left\|\sum_{t=1}^{T}\gamma_t\bar{\sigma}_{w_t}\right\|_{\infty}\right]$$

$$\leq \sqrt{2\ln(m)}\sqrt{\mathbb{E}_{\mathbb{P}2T}\left[\left\|\sum_{t=1}^{T}\gamma_t\bar{\sigma}_{w_t}\right\|_{\infty}^2\right]}$$

$$\leq \mathcal{O}(1)\mathcal{L}\ln^{1.5}(m)\sqrt{\sum_{t=1}^{T}\gamma_t^2/N}. \qquad (10.16)$$

We recall that $\Omega_V(d_{\Delta_m\times\Delta_n^M}) = \sqrt{2}$. Let us now assume that we run Algorithm 10.2 with constant step-sizes $\gamma > 0$ satisfying (10.9). Applying Theorem 4.4 together with (10.12), (10.13), (10.15), and (10.16), we obtain:

$$\mathbb{E}_{\mathbb{P}2T}\left[\max_{X\in\Delta_n^M}\phi(\bar{y},X)\right] - \phi^* \leq \mathbb{E}_{\mathbb{P}2T}\left[\max_{X\in\Delta_n^M}\phi(\bar{y},X) - \min_{y\in\Delta_m}\phi(y,\bar{X})\right] \leq \bar{\epsilon},$$

where $\bar{X} := (1/T)\sum_{t=1}^{T}\exp(\bar{V}_t)/\mathrm{Tr}[\exp(\bar{V}_t)]$ and

$$\bar{\epsilon} := \mathcal{O}(1)\left(\frac{1}{T\gamma} + \frac{\mathcal{L}\ln^{1.5}(m)}{\sqrt{NT}} + \frac{\mathcal{L}\ln(m)}{N} + \frac{\mathcal{L}^2\ln^2(m)\gamma}{N} + \frac{\mathcal{L}^2\ln^2(m)\gamma}{N^2}\right)$$

$$\leq \mathcal{O}(1)\left(\frac{1}{T\gamma} + \frac{\mathcal{L}\ln^{1.5}(m)}{\sqrt{NT}} + \frac{\mathcal{L}\ln(m)}{N} + \frac{\mathcal{L}^2\ln^2(m)\gamma}{N}\right).$$

Optimizing this last bound with respect to the step-size γ, we obtain:

$$\gamma^*(N) = \frac{1}{\mathcal{L}\ln(m)}\sqrt{N/T}.$$

In practice, it is attractive to have step-sizes that are as long as possible. Typically, the longer the step-sizes, the faster the algorithms in practice. We thus choose N such that $\gamma^*(N)$ is as large as possible while ensuring Condition (10.9). That is, we set

$$N = \left\lfloor\frac{T\ln(m)}{8\ln(n)}\right\rfloor \qquad \text{and} \qquad \gamma^* = \frac{1}{2\mathcal{L}\sqrt{2\ln(m)\ln(n)}}. \qquad (10.17)$$

We conclude:

$$\mathbb{E}_{\mathbb{P}2T}\left[\max_{X\in\Delta_n^M}\phi(\bar{y},X)\right] - \phi^* \leq \mathcal{O}(1)\frac{\mathcal{L}}{T}\left(\ln(m)\sqrt{\ln(n)} + \ln(n)\right). \qquad (10.18)$$

10.3.2.3 Arithmetic complexity

We are ready to give the arithmetic complexity of Algorithm 10.2. With K_t and N chosen as in (10.10) and (10.17), the computations of $\hat{g}_{\xi_{2t-1},K_t}(V_{t-1})$ and $\hat{g}_{\xi_{2t},K_t}(\bar{V}_t)$ require

$$\mathcal{O}\left(n^2 T^2 \sqrt{\ln(m)/\ln(n)} + mS\right)$$

arithmetic operations; see (10.7) for the definitions of $\hat{g}_{\xi_{2t-1},K_t}(V_{t-1})$ and $\hat{g}_{\xi_{2t},K_t}(\bar{V}_t)$. The remaining operations that we are supposed to carry out at iteration t of Algorithm 10.2 need in total $\mathcal{O}(n^2 + mS)$ arithmetic operations. Combining the iteration cost with (10.18), we observe that the following result holds.

Theorem 10.2 *Choose $\epsilon > 0$. Randomized Mirror-Prox (RMP) methods need on average*

$$\mathrm{Compl}_{\mathsf{RMP}}^{\mathrm{rand}}(\mathrm{DSP}, \epsilon) = \mathcal{O}\left(\frac{\mathcal{L}^3 n^2 \ln^{3.5}(m) \ln^{2.5}(n)}{\epsilon^3} + \frac{\mathcal{L} mS \ln(m) \ln(n)}{\epsilon}\right)$$

arithmetic operations to compute a point $\bar{y} \in \Delta_m$ for which $\max_{X \in \Delta_n^M} \phi(\bar{y}, X) - \phi^* \le \epsilon$.

Comparison: We conclude by comparing $\mathrm{Compl}_{\mathsf{RMP}}^{\mathrm{rand}}(\mathrm{DSP}, \epsilon)$ with the efficiency estimates $\mathrm{Compl}_{\mathsf{IP}}(\mathrm{PSP}, \epsilon)$ of Interior-Point methods, $\mathrm{Compl}_{\mathsf{RMD}}^{\mathrm{rand}}(\mathrm{DSP}, \epsilon)$ of randomized Mirror-Descent methods, and $\mathrm{Compl}_{\mathsf{ST}}(\mathrm{DSP}, \epsilon)$ of Smoothing Techniques. We refer to (8.12), Theorem 8.2, and Theorem 9.1 for the complexity results of the competitors. As usual, we suppose that $n \ge m$, choose $\beta \in [2,3]$ such that $n^\beta = mS$, and neglect all logarithmic terms in the complexity results. It holds that $\mathrm{Compl}_{\mathsf{RMP}}^{\mathrm{rand}}(\mathrm{DSP}, \epsilon) \ll \mathrm{Compl}_{\mathsf{IP}}(\mathrm{PSP}, \epsilon)$ if

$$\max\left\{\mathcal{L}^3/\epsilon^3, n^{\beta-2}\mathcal{L}/\epsilon\right\} \ll mn^{1.5},$$

that is, if

$$\mathcal{L}/\epsilon \ll m^{1/3} n^{1/2}.$$

We have $\mathrm{Compl}_{\mathsf{RMP}}^{\mathrm{rand}}(\mathrm{DSP}, \epsilon) \ll \mathrm{Compl}_{\mathsf{RMD}}^{\mathrm{rand}}(\mathrm{DSP}, \epsilon)$ if

$$\mathcal{L}/\epsilon \ll n^{\beta-2}.$$

Finally, the relation $\mathrm{Compl}_{\mathsf{RMP}}^{\mathrm{rand}}(\mathrm{DSP}, \epsilon) \ll \mathrm{Compl}_{\mathsf{ST}}(\mathrm{DSP}, \epsilon)$ is satisfied if

$$\mathcal{L}/\epsilon \ll n^{1/2} \qquad \text{and} \qquad 2 \le \beta \ll 3.$$

We conclude that $\mathrm{Compl}_{\mathrm{RMP}}^{\mathrm{rand}}(\mathrm{DSP}, \epsilon)$ is much smaller than

$$\min\left\{\mathrm{Compl}_{\mathrm{IP}}(\mathrm{PSP}, \epsilon), \mathrm{Compl}_{\mathrm{RMD}}^{\mathrm{rand}}(\mathrm{DSP}, \epsilon), \mathrm{Compl}_{\mathrm{ST}}(\mathrm{DSP}, \epsilon)\right\}$$

if

$$\mathcal{L}/\epsilon \ll \min\left\{n^{\beta-2}, n^{1/2}\right\} \qquad \text{and} \qquad 2 \le \beta \ll 3.$$

Application to (SDP): Note that we cannot derive any result on the arithmetic complexity of Algorithm 10.2 when used as basic method for solving (SDP). Theorem 10.2 concerns a feasible ϵ-solution from the set Δ_m and not a feasible ϵ-solution from the set Δ_n^M, which would be required by the transformation strategies described in Chapter 8.

Numerical results

IN this chapter, we present numerical results for Interior-Point methods, randomized Mirror-Descent schemes, (accelerated) Smoothing Techniques, and (randomized) Mirror-Prox methods when applied to randomly generated instances of Problem (DSP). These numerical results do not only reflect adequately our theoretical expectations concerning the performance of the methods, but they also prove the practical usefulness of both accelerated Smoothing Techniques and randomized Mirror-Prox methods. In the numerical experiments that we perform, these methods show the best performance for problems involving matrices of size 400×400 or larger. We obtain remarkably strong numerical results for accelerated Smoothing Techniques: they are able to solve a randomly generated instance of (DSP) involving 100 matrices of size $12'800 \times 12'800$ up to an absolute accuracy of 0.0012 in just less than four hours.

Contributions: The numerical results presented in this chapter are original.
Relevant literature: Parts of this chapter are taken from [BBN11].

11.1 Construction of the problem instances

We consider randomly generated instances of the problem:

$$\min_{y \in \Delta_m} \lambda_{\max}(\mathcal{A}(y)), \qquad \mathcal{A}(x) := \sum_{j=1}^{m} D_j y_j, \qquad D_j := j^{3/2} \bar{D}_j, \qquad (11.1)$$

where $\bar{D}_j$ are sparse symmetric $(n \times n)$-matrices for any $1 \leq j \leq m$, that is, we are confronted with instances of Problem (DSP). The matrices $\bar{D}_j$ are constructed as follows. We first generate the random symmetric matrix $\bar{D}_1$ using the Matlab built-in function *sprandsym(n,d)*, where n corresponds to the matrix size and d to the targeted fraction of non-zero entries in $\bar{D}_1$. In all our experiments, we set $d = 0.1$, that is, $\bar{D}_1$ is supposed to have about $S = n^2/10$ non-zero entries. Now, we construct matrices U_j that have the same sparsity pattern as $\bar{D}_1$ and whose non-zeros entries are drawn from the standard normal distribution using the Matlab built-in function *randn()*, where $2 \le j \le m$. In order to ensure symmetry of the matrices, we set $\bar{D}_j = (U_j + U_j^T)/2$ for any $2 \le j \le m$. We fix m as 100 in all numerical experiments that we perform.

Recall that we write $\mathcal{L}$ for the maximal absolute eigenvalue of all symmetric matrices D_j, where $1 \le j \le m$. We approximate this parameter by applying the Power method to the matrices D_j and taking the maximum of the computed values afterwards. In order to indicate clearly that we use an estimate of $\mathcal{L}$, we denote this approximation by $\mathcal{L}'$. We solve all instances of Problem (11.1) up to an accuracy of $\delta := \epsilon \mathcal{L}'$. In the experiments that we perform, the target accuracy ϵ is set to 0.002.

All numerical results that we present in this chapter are – if not specified differently – averaged over ten runs and are obtained on a computer with 24 processors, each of them with 2.67 GHz, and with 96 GB of RAM.

11.2 Implementation details of the methods

We solve random instances of (11.1) by Interior-Point methods, by randomized Mirror-Descent schemes, by Smoothing Techniques (with and without local L-estimation), and by Mirror-Prox methods (with and without randomized matrix exponential computations). In this section, we discuss the implementation of the different methods.

At every iteration, randomized Mirror-Descent schemes, Smoothing Techniques, and Mirror-Prox methods require the constant $\mathcal{L}$ to compute the next iterate. However, we do not have the exact value of $\mathcal{L}$, but its approximation $\mathcal{L}'$. Typically, this approximation $\mathcal{L}'$ given by the Power method underestimates the true value $\mathcal{L}$, which results in algorithms that are accelerated artificially and in a way that is theoretically forbidden. In order to be on the safe side, we use $\bar{\mathcal{L}} := 1.1 \cdot \mathcal{L}'$ for the computation of the next iterates in the aforementioned algorithms.

11.2.1 Interior-Point methods and randomized Mirror-Descent schemes

We utilize SeDuMi, an open-source Matlab implementation of an efficient **Interior-Point (IP) method**; see [Sed, Stu99] for more details on SeDuMi and for a free download of it. According to our experience, it is performance-wise much more favorable to apply SeDuMi to the dual problem of (11.1) instead of running it on its primal form. We recall that the dual of (11.1) takes the form

$$\max_{X \in \Delta_n^M} \min_{1 \leq j \leq m} \langle D_j, X \rangle_F,$$

which can be rewritten in a linear form; see Section (8.11) for the details. We apply SeDuMi with its standard accuracy, which is 10^{-8}, and not with the accuracy $\epsilon \mathcal{L}'$ as for the other methods. When running SeDuMi with the accuracy $\epsilon \mathcal{L}'$, we end up surprisingly with numerical troubles.

The implementation of **randomized Mirror-Descent (RMD) schemes** is due to Arkadi Nemirovski. All maximal eigenvalues and leading eigenvectors of symmetric matrices that are needed in the course of the algorithm are approximated by a randomized version of the Power method suggested (and also implemented) by Arkadi Nemirovski.

We apply these methods in a black box manner and do not elaborate more on their implementation.

11.2.2 (Accelerated) Smoothing Techniques

We use two versions of Smoothing Techniques, namely:

1. **original Smoothing Techniques (ST)**, where we set $L_t = L_\mu$ for all $t \geq 0$ in Algorithm 4.1;

2. **accelerated Smoothing Techniques (AST)**, where $L_0 := L_\mu$ and

$$L_t := \frac{2}{\|u_t - x_t\|_1^2} \left(\overline{\phi}_\mu(u_t) - \overline{\phi}_\mu(x_t) - \langle \nabla \overline{\phi}_\mu(x_t), u_t - x_t \rangle \right)$$

for any $t \geq 1$ in Algorithm 4.1.

The smoothness parameter μ and the resulting Lipschitz constant L_μ are set as described in Section 9.2, that is,

$$\mu = \frac{\epsilon \bar{\mathcal{L}}}{2 \ln(n)} \qquad \text{and} \qquad L_\mu = \frac{2 \bar{\mathcal{L}} \ln(n)}{\epsilon}.$$

Given a pair $(\bar{y}, \bar{X}) \in \Delta_m \times \Delta_n^M$ of a primal and a dual feasible solution, we can compute the corresponding duality gap

$$\lambda_{\max}(\mathcal{A}(\bar{y})) - \min_{1 \leq j \leq m} \langle D_j, \bar{X} \rangle_F, \tag{11.2}$$

which we use as stopping criterion for both the original and the accelerated implementation of Smoothing Techniques. The first term is approximated by

N	CPU time [sec]		# iterations		CPU time per iteration [sec/iteration]
	mean	std	mean	std	
1	52	6	2'626	300	0.020
3	55	8	2'633	363	0.021
5	55	4	2'620	193	0.021
10	56	5	2'650	212	0.021
50	61	8	2'710	360	0.023
100	63	6	2'631	248	0.024
500	94	10	2'640	276	0.036
1000	117	11	2'635	256	0.044
5000	360	41	2'634	257	0.147

Table 11.1: *CPU time (mean, standard deviation), number of iterations (mean, standard deviation), and average CPU time per iteration required by **randomized Mirror-Prox methods** with different samples sizes N for solving random instances of Problem (11.1) involving a hundred matrices of size* 100×100.

a Arkadi Nemirovski's randomized version of the Power method. When the first approximation obtained by the randomized version of the Power method yields to a value that is smaller than $\epsilon \mathcal{L}'$, we recompute the duality gap using the Matlab built-in functions *max()* and *eig()* and terminate only if this new value fulfills the stopping condition. In both implementations of Smoothing Techniques, we check the above stopping criterion at every 100-th iteration. Additionally for accelerated Smoothing Techniques, we verify this condition at every of the first hundred iterations.

11.2.3 (Randomized) Mirror-Prox methods

We run three different implementations of Mirror-Prox methods, which are:

1. **original Mirror-Prox (MP) methods** (Algorithm 10.1) with constant step-sizes that are as long as possible, that is, with step-sizes

$$\gamma_t = \frac{1}{2\bar{\mathcal{L}}\sqrt{\ln(m)\ln(n)}} \qquad \forall\, t \geq 1.$$

We use the same stopping criterion as for Smoothing Techniques and verify it at every 100-th iteration;

2. **decelerated Mirror-Prox (DMP) methods** (Algorithm 10.1) with the same step-sizes as in randomized Mirror-Prox methods, that is, with step-sizes

$$\gamma_t = \frac{1}{2\bar{\mathcal{L}}\sqrt{2\ln(m)\ln(n)}} \qquad \forall\, t \geq 1.$$

The stopping criterion and its checking procedure are as above;

3. **randomized Mirror-Prox (RMP) methods** with constant step-sizes

$$\gamma_t = \frac{1}{2\bar{\mathcal{L}}\sqrt{2\ln(m)\ln(n)}} \qquad \forall\, t \geq 1;$$

see Algorithm 10.2 and Equation (10.17). Given a matrix $W \in \mathcal{S}_n$, we choose the truncation level K_W of the matrix exponential Taylor series approximation in Algorithm 10.2 according to the following formula:

$$K_W = \left\lfloor \max\left\{ \log(1/\rho), \exp(1)\,\|W\|_{(\infty)} \right\} \right\rceil, \qquad \rho := 10^{-3}.$$

Note that this setting slightly deviates from the truncation level derived in Proposition 10.2, where the second quantity in the max-expression is multiplied by a factor of $\exp(1)$. The ∞-norm of W is computed approximately using the Power method (and corrected in hindsight by a factor of 1.2).

In accordance to (10.17) and (10.18), we need to choose the sample size N as:

$$N = \frac{\mathcal{O}(1)\ln^2(m)}{\epsilon\sqrt{\ln(n)}}. \tag{11.3}$$

In Table 11.1, we give the CPU time (mean, standard deviation) and the number of iterations (mean, standard deviation) needed to find a solution with an accuracy of $\epsilon\mathcal{L}'$, as well as the average CPU time per iteration for different samples sizes N. We observe that the smaller the sample size, the lower the CPU time that is required to approximately solve the problem instances. Surprisingly, we can choose a very small sample size without sacrificing any iterations. Let us illustrate this observation with an example. According to (11.3) and with an absolute constant of 1, we are supposed to choose N as about 5000. With this parameter choice, we need an average CPU time of 360 seconds. Using only one sample for each matrix exponential approximation, we can reduce the average CPU time by 85.6%. For the subsequent tests, we will thus choose one as sample size for every matrix exponential approximation.

	mean				std			
n	100	200	400	800	100	200	400	800
S	953	3'813	15'244	60'925	10	17	32	64
$\mathcal{L}'$	4'910	6'463	8'680	12'031	194	216	119	112

Table 11.2: *Number of non-zero entries in any D_j, $1 \leq j \leq m$, and the constant $\mathcal{L}'$ for different matrix sizes n. We show both mean and standard deviation.*

Consider the stopping criterion defined in (11.2). At any iteration $t \geq 1$ of randomized Mirror-Prox methods, the pair $(\bar{y}, \bar{X})$ of this definition corresponds to the average

$$\frac{1}{t} \sum_{\tau=1}^{t} \gamma_\tau (\bar{y}_\tau, \hat{\mathcal{E}}_{\xi_{2\tau}, K_\tau}(\bar{V}_\tau)),$$

where $(\bar{y}_\tau, \bar{V}_\tau)$ is defined in Algorithm 10.2 and $\hat{\mathcal{E}}_{\xi_{2\tau}, K_\tau}(\bar{V}_\tau)$ is specified in Equation (10.8) for any $1 \leq \tau \leq t$. In principle, the criterion (11.2) gives theoretically a desirable solution only if we use exact scaled exponentials instead of $\hat{\mathcal{E}}_{\xi_{2\tau}, K_\tau}(\bar{V}_\tau)$; see Theorem 4.4. Nevertheless, $\hat{\mathcal{E}}_{\xi_{2\tau}, K_\tau}(\bar{V}_\tau)$ is in the matrix simplex by construction, and the number of terms we use in the Taylor exponential is large enough to justify a very accurate approximation, so that $\bar{X}$ can be considered as an adequate approximate solution to our problem. We implement the same duality gap checking procedure as for original Smoothing Techniques.

11.3 Comparison of the methods

In order to compare the performance of the different methods, we run them on random instances of (11.1) for different matrix sizes, namely for sizes $n = 100, 200, 400, 800$. For every problem size, we generate ten problem instances and average the performance of each method on these instances afterwards.

Table 11.2 shows the number of non-zero entries (mean, standard deviation) of the matrices D_j and the constant $\mathcal{L}'$ (mean, standard deviation) for each problem size. Recall that $\mathcal{L}'$ denotes the maximal absolute eigenvalue of all matrices D_j computed approximately by the Power method.

Let us turn our attention now to Table 11.3, where we show the CPU time (mean, standard deviation), the number of iterations needed in practice

CPU time [sec]

	mean				std			
n	100	200	400	800	100	200	400	800
IP	4	42	885	$*\,*\,*$	<1	<1	53	$*\,*\,*$
RMD	153	424	$1'624$	$6'160$	17	13	190	795
ST	75	265	$1'258$	$5'511$	2	4	81	225
AST	22	3	7	26	19	<1	<1	1
MP	46	130	541	$2'507$	5	7	20	251
DMP	65	183	760	$3'469$	6	10	29	319
RMP	53	149	529	$2'000$	4	9	25	55

 # required iterations in practice

	mean				std			
n	100	200	400	800	100	200	400	800
IP	29	28	28	$*\,*\,*$	1	2	2	$*\,*\,*$
RMD	$17'507$	$18'166$	$18'423$	$18'915$	$1'835$	593	656	512
ST	$6'060$	$6'640$	$7'010$	$7'430$	151	52	74	48
AST	$1'003$	17	13	12	911	2	<1	<1
MP	$1'840$	$1'724$	$1'760$	$1'810$	171	87	52	32
DMP	$2'600$	$2'410$	$2'470$	$2'540$	236	129	68	70
RMP	$2'648$	$2'460$	$2'470$	$2'414$	200	143	106	64

 # required iterations in theory

n	100	200	400	800
ST	$10'131$	$10'867$	$11'556$	$12'206$
MP	$5'066$	$5'434$	$5'778$	$6'103$
DMP	$7'164$	$7'684$	$8'171$	$8'631$

average CPU time / per iteration [sec/iterations]

n	100	200	400	800
RMD	0.009	0.023	0.088	0.326
ST	0.012	0.040	0.179	0.742
AST	0.022	0.149	0.540	2.146
MP	0.025	0.076	0.308	1.385
DMP	0.025	0.076	0.308	1.366
RMP	0.020	0.061	0.214	0.829

Table 11.3: *CPU time (mean, standard deviation), number of iterations needed in practice (mean, standard deviation), and the number of iterations that is theoretically needed by the different methods to find a solution to Problem (11.1) with accuracy $\epsilon \mathcal{L}'$. In the last table, we plot the average CPU time per iteration. The notation "$*\,*\,*$" means that the method did not finish within four hours.*

 # terms in Taylor series approximation

n	average				std			
	100	200	400	800	100	200	400	800
RMP	8.8	9.4	10.0	10.4	0.3	0.2	0.1	< 0.1

Acceleration of L in Algorithm 4.1
(ratios need to be multiplied with 10^{-6})

n	average				std			
	100	200	400	800	100	200	400	800
AST	$9.2 \cdot 10^4$	6.7	2.5	1.5	$9.2 \cdot 10^4$	2.2	0.6	0.2

Table 11.4: *Upper table: Number of terms (mean, standard deviation) of the matrix exponential Taylor series approximation considered in **randomized Mirror-Prox methods**. Lower table: Ratio (11.4) for **accelerated Smoothing Techniques**.*

(mean, standard deviation), and the number of iterations that is theoretically required by the different methods to find a solution to Problem (11.1) with accuracy $\epsilon \mathcal{L}'$. Note that the theoretical number of iterations is deterministic due to the fact that the solution accuracy $\epsilon \mathcal{L}'$ depends on $\mathcal{L}'$. In addition, we present in the same table the average CPU time that is needed per iteration. When looking at the results presented in this table, we make the following observations:

⋄ For $n = 100$, Interior-Point (IP) methods require the least CPU time to compute an approximate solution. However, the CPU time is growing faster for these methods than for all other algorithms when we increase n. For $n = 800$, Interior-Point methods did – as single algorithm – not manage to derive a solution within four hours. We terminated the algorithm after this time span. These numerical results reflect our theoretical expectations regarding the fast growth of the running time for Interior-Point methods with respect to n.

⋄ For $n = 200, 400, 800$, accelerated Smoothing Techniques (AST) show by far the best performance. On average, they need only 3 (for $n = 200$), 7 (for $n = 400$), and 26 ($n = 800$) seconds to solve these problems approximately. Randomized Mirror-Prox (RMD) methods constitute the second best algorithm for both $n = 400$ and $n = 800$. They require on average 529 and $2'000$ seconds, respectively. These proves the practical relevance of accelerated Smoothing Techniques and randomized Mirror-Prox methods when compared to the existing methods. Still, note the dramatic difference between the performance of randomized

Mirror-Prox methods and accelerated Smoothing Techniques. The later procedure requires on average about 1.3% of the CPU time of randomized Mirror-Prox methods if $n = 800$.

◇ The strong performance of accelerated Smoothing Techniques (AST) has its reason in the remarkably small numbers of iterations that are required to find approximate solutions. For $n = 800$, they need on average only twelve iterations, which corresponds to about 0.2% of the number of iterations needed by original Smoothing Techniques. Remarkably, the number of required iterations is even decaying when we increase n.

◇ We recall that decelerated Mirror-Prox (DMP) methods use the same step-sizes as randomized Mirror-Prox (RMP) schemes. When we compare the average CPU time needed by these methods, we observe that randomized Mirror-Prox schemes need only about 81.5%, 81.4%, 69.6%, and 57.7% of the CPU time of the decelerated version (for n equal to 100, 200, 400, and 800, respectively). Remarkably enough, the number of iterations required by these methods in practice is roughly the same. The differences in the total running time are due to the lower iteration cost of randomized Mirror-Prox methods.

◇ The theoretical number of iterations of original Mirror-Prox (MP) methods and of its decelerated version (DMP) deviates by factor of $\sqrt{2}$, that is, of about 1.414, which is due to the different step-sizes. Interestingly, we observe the same number for the deviation of the iteration numbers that are required in practice: we obtain ratios that lie on average in the interval $[1.398, 1.413]$. That is, the speed of a Mirror-Prox methods scales proportionally with the length of the step-sizes: the longer the step-sizes, the faster the method in practice.

◇ Smoothing Techniques (ST) have a lower iteration cost than original Mirror-Prox (MP) methods, as they require the computation of only one instead of two matrix exponentials at every step.

◇ Among all First-Order methods, randomized Mirror-Descent (RMD) schemes show the weakest performance regarding the average CPU time, because of the relatively high numbers of iterations that are needed by them. They require about 2.5 to 10.5 times more iterations than standard advanced First-order methods such Smoothing Techniques and Mirror-Prox algorithms. From theory, we indeed estimate this behavior as the iteration count in Mirror-Descent methods grows one order of magnitude faster with respect to the accuracy than in advanced First-Order schemes. In addition, we observe that randomized

Mirror-Descent methods require the least CPU time per iteration. We recall that we need here to approximate only the largest eigenvalue and the leading eigenvector. This is in contrast to all other First-Order methods, where we are supposed to compute or to approximate the exponential of at least one symmetric matrix.

In Table 11.4, we express the number of terms that we consider in the Taylor series approximation (mean and standard deviation) in randomized Mirror-Prox schemes. We observe that we need to take into account only a very small number, namely about ten of them.

In the same table, we show for accelerated Smoothing Techniques the ratio

$$\frac{\sum_{t=1}^{T} L_t}{L_0 T},\tag{11.4}$$

where we use the notation of Algorithm 4.1. We recognize that the local estimates are significantly smaller than the global Lipschitz constant: on average, the above ratio takes a value between $1.5 \cdot 10^{-6}$ and $9.2 \cdot 10^{-2}$.

11.4 Solving very large-scale problems

We apply now accelerated Smoothing Techniques to very large-scale instances of Problem (11.1). The numerical results that we present in this section are obtained by single runs.

As shown in Table 11.5, we need $22'300$ seconds, that is, about six hours and twelve minutes, to solve approximately a random instance of (11.1) with matrices of size $12'800 \times 12'800$ and with about $15.6 \cdot 10^6$ non-zero entries. Target accuracy and number of matrices D_j are chosen as in the previous section. Interestingly, the number of iterations that are carried out is always about twelve, independent of the problem size.

In the same table, we present some additional numerical results for accelerated Smoothing Techniques when applied to (11.1), but where we use matrices $D_j = \bar{D}_j/\sqrt{n}$ instead of $j^{3/2} D_j$ for any $1 \le j \le m$. In contrast to the examples that we considered before, the parameter $\mathcal{L}'$ is not increasing any more with n, but stays more or less constant; see [Wig58] for a justification of this effect. The number of matrices that we consider, the sparsity structure, and the definition of the target solution accuracy remain unchanged. We observe that we need just less than four hours for solving the randomly generated instance involving matrices of size $12'800 \times 12'800$ up to an accuracy of $\epsilon \mathcal{L}' = 0.0012$. As for the previous example, only a very small number of iterations is required to find this approximate solution.

$$D_j = j^{3/2}\bar{D}_j$$

n	800	1'600	3'200	6'400	12'800
S	61'041	243'734	974'868	3'898'844	15'591'553
$\mathcal{L}'$	11'788	16'919	23'647	33'303	47'048
CPU time [sec]	29	140	654	3'414	22'300
# steps carried out	13	12	12	12	12
CPU time / step	2.2	11.7	54.5	284.5	1'858.3

$$D_j = \bar{D}_j/\sqrt{n}$$

n	800	1'600	3'200	6'400	12'800
S	61'087	243'660	974'349	3'897'199	15'593'629
$\mathcal{L}'$	0.605	0.610	0.602	0.604	0.604
CPU time [sec]	50	113	437	2'103	14'351
# steps carried out	22	10	8	7	7
CPU time / step	2.3	11.3	54.6	300.4	2'050.1

Table 11.5: *Upper table: CPU time, number of required steps, and CPU time per iteration required by **accelerated Smoothing Techniques** when applied to very large-scale instances of Problem (11.1). Lower table: Same data, but with a different generation of the matrices D_j in (11.1). We use $D_j = \bar{D}_j/\sqrt{n}$ instead of $j^{3/2}\bar{D}_j$ for any $1 \le j \le m$.*

Chapter 12

Conclusions and outlook

IN this chapter, we give a review of the results presented in this dissertation and perform an outlook of possible future work.

12.1 Conclusions

This thesis aimed at developing methods for solving (structured) large-scale semidefinite optimization problems approximately. Theoretically, these problems can be solved by polynomial-time Interior-Point methods. These methods are an excellent tool for finding solutions with a low approximation error to medium-scale semidefinite optimization problems, as this quantity enters the complexity result through the logarithm. However, the worst-case complexity estimate of Interior-Point methods grows with the power 3.5 in both the matrix size and the number of constraints, which hampers the resolution of large-scale problems in practice.

In this thesis, we presented alternative approaches to solve slightly structured large-scale semidefinite optimization problems up to a moderate accuracy. The algorithmic foundation of these approaches is formed by advanced First-Order schemes such as Smoothing Techniques and Mirror-Prox methods.

These methods make very specific structural requirements on the problem. However, semidefinite optimization problems do not fit in this framework, given their general form. In a first step, we discussed two problem transformations that allowed us to rewrite slightly structured semidefinite optimization problems as matrix saddle-point problems to which the advanced First-Order methods are applicable. The first transformation is – to the best of our knowledge – new and can be applied to problems whose linear objective function is defined by a positive definite matrix. The second transformation scheme is known and requires a scaling constraint. This problem transformation is particularly favorable in the sense that the sparsity of the input matrices is transferred to the matrix saddle-point problem.

Therefore, solving these structured semidefinite optimization problems reduces to the resolution of matrix saddle-point problems. It is known that these matrix saddle-point problems can be approximately solved by Interior-Point methods, by (randomized) Mirror-Descent methods, and by advanced First-Order methods such as Smoothing Techniques and Mirror-Prox methods. In Table 12.1, we give an overview of the complexity results of these methods. Interior-Point methods are particularly tailored for solving medium-scale problems up to a high accuracy. Randomized Mirror-Descent schemes are at the other end of the landscape and can be applied to very large-scale problems, provided that we require only a very low solution accuracy. The iteration count of these methods grows with the order $\mathcal{O}(1/\epsilon^2)$ with respect to the accuracy $\epsilon > 0$. In between, we have advanced First-Order methods such as Smoothing Techniques and Mirror-Prox methods, whose iteration count grows one order of magnitude slower than for Mirror-Descent schemes.

By applying Smoothing Techniques or Mirror-Prox methods to these matrix saddle-point problems, we developed a procedure for solving slightly structured large-scale semidefinite optimization problems. The iteration count of the resulting schemes grows linearly in both the inverse of the accuracy and in the number of constraints. Similar results were obtained independently and simultaneously by Iyengar et al. [IPS05, IPS11].

At every iteration of these methods, the computation of one ore two matrix exponentials is needed. Standardly, this operation is performed through an eigendecomposition of the symmetric $(n \times n)$-matrix, requiring $\mathcal{O}(n^3)$ arithmetic operations and representing the cost critical computation for large-scale problems. In a joint project with Arkadi Nemirovski, we developed and investigated both theoretically and numerically an alternative computation of matrix exponentials. Instead of the true value, we used a random approximation of the matrix exponential in Mirror-Prox methods. This random approximation is based on a vector sampling and on an appropriate truncation of the matrix exponential Taylor series. We verified that there exists a reasonable range for the problem parameters where this method outperforms all its competitors with respect to the worst-case complexity estimates. We noted that sparsity of the symmetric matrix, whose exponential we need to compute, is in full favor of this randomized method. While sparsity cannot be exploited in the standard matrix exponentiation technique, we may reduce the computation cost of the randomized approximation in accordance to the

Solution method	Complexity
Interior-Point methods	$\tilde{\mathcal{O}}\left(\sqrt{m+n}\left[mn^3 + mS + m^3\right]\right)$
Smoothing Techniques	$\tilde{\mathcal{O}}\left(\mathcal{L}[n^3 + mS]/\epsilon\right)$
Mirror-Prox methods	$\tilde{\mathcal{O}}\left(\mathcal{L}[n^3 + mS]/\epsilon\right)$
Randomized Mirror-Descent methods	$\tilde{\mathcal{O}}\left(\mathcal{L}^{2.5}n^2/\epsilon^{2.5} + \mathcal{L}^2 mS/\epsilon^2\right)$
Randomized Mirror-Prox methods	$\tilde{\mathcal{O}}\left(\mathcal{L}^3 n^2/\epsilon^3 + mS/\epsilon\right)$

Table 12.1: *Rough complexity estimates of different methods when applied to matrix saddle-point problems; see (DSP) for the problem definition. We use the $\tilde{\mathcal{O}}$-notation to indicate that all logarithmic terms are suppressed. We denote by m, n, $\mathcal{L}$, S, and ϵ the number of input matrices, the matrix size, the maximal absolute eigenvalue of all input matrices, the maximal number of non-zero entries in the input matrices, and the target accuracy.*

degree of sparsity. Numerical results showed that the randomized Mirror-Prox method requires only about 58% of the CPU time that was needed by its deterministic counterpart for solving approximately randomly generated instances of matrix saddle-point problems with a hundred matrices of size 800×800. These matrices had the same sparsity pattern, each of them with about $60'000$ non-zero entries. The relevant theoretical results in this joint project were mainly due to Arkadi Nemirovski. We were concerned mostly with the design and the performance of numerical tests.

In this thesis, we presented a refinement of Smoothing Techniques, or, more precisely, of one of the two ingredients of Smoothing Techniques. Smoothing Techniques are a two-stage procedure. In a first step, a smooth approximation of the non-differentiable objective function is formed. In a second step, an optimal First-Order method is applied to the auxiliary problem. At every iteration of this optimal First-Order method, the Lipschitz constant of the gradient of the objective function is used. We introduced a new version of this optimal First-Order method where we replaced this global constant by local estimates. The theoretical worst-case complexity estimate for the refined First-Order method is of the same order as for the vanilla scheme. More interestingly, we observed that this refined method requires significantly less iterations than its original counterpart for finding approximate solutions in practice. By integrating this refined version of the method in Smoothing Techniques, we could solve a randomly generated instance of a matrix saddle-point problem involving a hundred matrices of size $12'800 \times 12'800$ up to an accuracy of 0.0012 in just less than four hours.

As a side result of this thesis, we obtained a complete and consistent embedding of the Hedge algorithm in the context of Dual Averaging schemes. It was known that the Hedge algorithm can be interpreted as a Mirror-Descent scheme. However, this interpretation is – as we showed in this thesis – not consistent with the setting of the Hedge algorithm. Interpreting the Hedge algorithm as a Dual Averaging scheme, we were able to eliminate these inconsistencies. Based on the insights gained from this new perspective on the method, we developed three new versions of the Hedge algorithm with convergence guarantees that are better or at least as good as the convergence result for the vanilla scheme. Numerical results showed that all these modified methods perform better than the vanilla algorithm in practice.

12.2 Outlook

As the numerical results showed, only a small number of iterations of the refined optimal First-Order method in Smoothing Techniques is needed to solve very large-scale matrix saddle-point problems approximately. At every iteration of this method, an exponential of a symmetric matrix needs to be computed. When using standard techniques to perform this operation, we hamper the resolution of problems with a huge scale. As a next step, it would be very natural to apply similar techniques as we did for Mirror-Prox methods to Smoothing Techniques to accelerate this critical operation.

All the First-Order methods studied in this thesis rely on (a slightly weaker version of) Assumption 2.1. That is, these methods are only effective if we have access to a distance-generating function for which can easily compute projections onto the feasible set. Unfortunately, we know such distance-generating functions only for a very restricted family of feasible sets; for instance, we can utilize the shifted negative matrix entropy function for the matrix simplex. For a broader applicability of these methods, the discovery of new distance-generating functions that allow us an efficient computation of projections onto alternative feasible sets is crucial.

Finally, it is known that AdaBoost is based on the Hedge algorithm. However, this algorithm is much more prominent than the method it is based on. In this thesis, we showed that Hedge algorithm can be recast as a Dual Averaging scheme. Having a similar result for AdaBoost would be of great interest.

Appendix **A**

Regularity of norms

$\mathbf{I}$N this chapter, we review the notion of regular norms (and spaces) and give some examples. Of particular interest for this thesis, regularity of norms is used to install two results that are used in Section 10.3.2 and for the proof of Proposition 10.1. This review follows the working paper [Nem04b]. Parts of this review chapter are taken from [BBN11].

Let us start with the first definition. We equip $\mathbb{R}^n$ with a norm $\|\cdot\|$ and choose $\kappa \geq 1$.

Definition A.1 (κ-smooth) *We say that the space $(\mathbb{R}^n, \|\cdot\|)$ is κ-smooth if the function $h : \mathbb{R}^n \to \mathbb{R}_{\geq 0} : x \mapsto \|x\|^2$ is continuously differentiable and if*

$$h(x + y) \leq h(x) + \langle \nabla h(x), y \rangle + \kappa h(y) \qquad \forall\, x, y \in \mathbb{R}^n.$$

Example A.1 (Proof of Example 2.1 in [Nem04b]) *If $2 \leq p < \infty$ and $n \geq 3$, the space $(\mathbb{R}^n, \|\cdot\|_p)$ is $(p - 1)$-smooth.*

Of particular interest in this thesis is the space $(\mathbb{R}^n, \|\cdot\|_\infty)$, which is not κ-smooth, but "almost" so. Let us clarify now what we mean by "almost κ-smooth".

Definition A.2 (κ-regular) *Both the space $(\mathbb{R}^n, \|\cdot\|)$ and the norm $\|\cdot\|$ are called κ-regular, if there exist $\kappa_+ \in [1, \kappa]$ and a norm $\|\cdot\|_+$ on $\mathbb{R}^n$ with the following two characteristics:*

1. the space $(\mathbb{R}^n, \|\cdot\|_+)$ is κ_+-smooth;

2. we have

$$\|x\|^2 \le \|x\|_+^2 \le \frac{\kappa}{\kappa_+}\|x\|^2 \qquad \forall\, x \in \mathbb{R}^n.$$

Definition A.3 (Regularity constant) *The regularity constant $\kappa(\mathbb{R}^n, \|\cdot\|)$ of $(\mathbb{R}^n, \|\cdot\|)$ is defined as:*

$$\kappa(\mathbb{R}^n, \|\cdot\|) := \inf\{\kappa \ge 1 : (\mathbb{R}^n, \|\cdot\|) \text{ is } \kappa\text{-regular}\}.$$

Example A.2 (Example 2.1 in [Nem04b]) *For $2 \le p \le \infty$ and $n \ge 3$, we have:*

$$\kappa(\mathbb{R}^n, \|\cdot\|_p) \le \min\{p - 1, 2\ln(n)\}.$$

From now on, assume that $(\mathbb{R}^n, \|\cdot\|)$ has a regularity constant of $\kappa(\mathbb{R}^n, \|\cdot\|)$. In addition, let $(\Xi, \mathcal{B}, \mathbb{P})$ be a Borel probability space and $T \in \mathbb{N}$. We consider T random vectors $\xi_t : \Xi \to \mathbb{R}^n$ and assume that they form a martingale difference sequence, that is:

$$\mathbb{E}_{\mathbb{P}}[\xi_1] = 0 \qquad \text{and} \qquad \mathbb{E}_{\mathbb{P}}\left[\xi_t | \xi_{[t-1]}\right] = 0 \qquad \forall\, 2 \le t \le T,$$

where we write $\xi_{[t-1]}$ for the subsequence $\xi_{t-1}, \ldots, \xi_1$.
The following result is used in Section 10.3.2.

Proposition A.1 (Proposition 3.1 in [Nem04b]) *If there exist $\sigma_1, \ldots, \sigma_T$ such that*

$$\mathbb{E}_{\mathbb{P}}\left[\|\xi_t\|^2\right] \le \sigma_t^2 < \infty \qquad \forall\, 1 \le t \le T,$$

we have:

$$\mathbb{E}_{\mathbb{P}^T}\left[\left\|\sum_{t=1}^{T} \xi_t\right\|^2\right] \le \kappa(\mathbb{R}^n, \|\cdot\|)\sum_{t=1}^{T} \sigma_t^2.$$

The proof of Proposition 10.1 is based on the following result on large deviations of random sums.

Theorem A.1 (Theorem 3.5 in [Nem04b]) *Choose $\chi \in (0, 2]$ and reals $\sigma_1, \ldots, \sigma_T > 0$ such that:*

$$\mathbb{E}_{\mathbb{P}}\left[\exp\left(\|\xi_t\|^\chi / \sigma_t^\chi\right) | \xi_{[t-1]}\right] \le \exp(1) \qquad \forall\, t = 1, \ldots, T.$$

(a) For all $c \ge 0$, we have:

$$\mathbb{P}\left[\left\|\sum_{t=1}^{T} \xi_t\right\| > c\sqrt{\kappa(\mathbb{R}^n, \|\cdot\|)\sum_{t=1}^{T} \sigma_t^2}\right] \le C_\chi \exp\left(-c^\chi / C_\chi\right),$$

where $C_\chi \ge 2$ is a constant that only depends on χ and that is continuous in χ.

(b) *With a properly chosen constant $c_\chi > 0$ that only depends on χ and that is continuous in χ, we have:*

$$\mathbb{E}_{\mathbb{P}^T}\left[\exp\left(\left(\frac{\left\|\sum_{t=1}^{T}\xi_t\right\|}{c_\chi\sqrt{\kappa(\mathbb{R}^n,\|\cdot\|)\sum_{t=1}^{T}\sigma_t^2}}\right)^{\chi}\right)\right] \leq \exp(1).$$

Appendix B

Proofs

$\mathbf{I}$N this chapter, we give the proofs of Theorem 3.5, Theorem 7.2, and Proposition 10.1. The proof of Proposition of 10.1 is taken from [BBN11].

B.1 Proof of Theorem 3.5

Choose $T \in \mathbb{N}_0$ and let the sequences $(x_t)_{t=0}^{T+1}$, $(u_t)_{t=0}^{T+1}$, $(z_t)_{t=0}^{T}$, $(\hat{x}_t)_{t=1}^{T+1}$, $(\gamma_t)_{t=0}^{T+1}$, $(\Gamma_t)_{t=0}^{T+1}$, $(\tau_t)_{t=0}^{T}$, and $(L_t)_{t=0}^{T}$ be generated by Algorithm 3.4. Recall that Inequality $(\mathcal{I}_t)$ holds for $0 \le t \le T$ if

$$\Gamma_t f(u_t) + \sum_{k=0}^{t-1} (L_{k+1} - L_k) \left(d(z_{k+1}) - \frac{1}{2} \| z_k - \hat{x}_{k+1} \|^2 \right) \le \psi_t, \qquad (\mathcal{I}_t)$$

where

$$\psi_t := \min_{x \in Q} \left\{ \sum_{k=0}^{t} \gamma_k \left(f(x_k) + \langle \nabla f(x_k), x - x_k \rangle \right) + L_t d(x) \right\}.$$

By its definition (see Algorithm 3.4), the element $z_t \in Q$ is the minimizer to the above optimization problem, which allows us to rewrite ψ_t as:

$$\psi_t = \sum_{k=0}^{t} \gamma_k \left(f(x_k) + \langle \nabla f(x_k), z_t - x_k \rangle \right) + L_t d(z_t).$$

We show by induction that Inequality $(\mathcal{I}_t)$ holds for any $0 \le t \le T$.

Lemma B.1 *Inequality $(\mathcal{I}_0)$ holds, that is, we have $\gamma_0 f(u_0) \leq \psi_0$.*

Proof: We apply the definition of u_0 (see Algorithm 3.4), Inequality (2.12), the condition on γ_0 saying that $\gamma_0 \in (0,1]$, and Theorem 2.10 in order to justify the following relations:

$$
\begin{aligned}
\psi_0 \; &:= \; \min_{x \in Q} \left\{ \gamma_0 \left(f(x_0) + \langle \nabla f(x_0, x - x_0) \rangle \right) + L_0 d(x) \right\} \\
&= \; \gamma_0 \left(f(x_0) + \langle \nabla f(x_0), u_0 - x_0 \rangle \right) + L_0 d(u_0) \\
&\geq \; \gamma_0 \left(f(x_0) + \langle \nabla f(x_0), u_0 - x_0 \rangle \right) + \frac{L_0}{2} \left\| u_0 - x_0 \right\|^2 \\
&\geq \; \gamma_0 \left(f(x_0) + \langle \nabla f(x_0), u_0 - x_0 \rangle + \frac{L_0}{2} \left\| u_0 - x_0 \right\|^2 \right) \\
&\geq \; \gamma_0 f(u_0).
\end{aligned}
$$

$\blacksquare$

Let us verify the inductive step.

Lemma B.2 *Let $0 \leq t \leq T - 1$. If Inequality $(\mathcal{I}_t)$ holds, also $(\mathcal{I}_{t+1})$ is true.*

Proof: Let $0 \leq t \leq T-1$ and assume that $(\mathcal{I}_t)$ holds. We make the following two definitions:

$$
\begin{aligned}
\chi_t \; &:= \; \sum_{k=0}^{t-1} (L_{k+1} - L_k) \left(d(z_{k+1}) - \frac{1}{2} \left\| z_k - \hat{x}_{k+1} \right\|^2 \right) \in \mathbb{R}, \\
s_t \; &:= \; \sum_{k=0}^{t} \gamma_k \nabla f(x_k) \in \mathbb{R}^n.
\end{aligned}
$$

In addition, we define the linear function:

$$
l_t : Q \to \mathbb{R} : x \mapsto l_t(x) = \sum_{k=0}^{t} \gamma_k \left(f(x_k) + \langle \nabla f(x_k), x - x_k \rangle \right).
$$

Choose $x \in Q$. The definition of z_t implies:

$$
0 \leq \left\langle L_t \nabla d(z_t) + \sum_{k=0}^{t} \gamma_k \nabla f(x_k), x - z_t \right\rangle = \langle L_t \nabla d(z_t) + s_t, x - z_t \rangle . \quad \text{(B.1)}
$$

As the Inequality $(\mathcal{I}_t)$ holds and as the function f is convex, we have:

$$
\psi_t \; \geq \; \Gamma_t f(u_t) + \chi_t \geq \Gamma_t \left(f(x_{t+1}) + \langle \nabla f(x_{t+1}), u_t - x_{t+1} \rangle \right) + \chi_t.
$$

This implies:

$$\psi_t + \gamma_{t+1}\left(f(x_{t+1}) + \langle \nabla f(x_{t+1}), x - x_{t+1}\rangle\right)$$
$$\geq \quad \Gamma_{t+1}f(x_{t+1}) + \gamma_{t+1}\langle \nabla f(x_{t+1}), x - z_t\rangle + \chi_t,$$

where we use the relations $\Gamma_{t+1} = \Gamma_t + \gamma_{t+1}$ and

$$
\begin{aligned}
\Gamma_t(u_t - x_{t+1}) &+ \gamma_{t+1}(x - x_{t+1}) \\
&= \quad \Gamma_t u_t - \Gamma_{t+1}x_{t+1} + \gamma_{t+1}x \\
&= \quad \Gamma_t u_t - \Gamma_{t+1}\left(\tau_t z_t + (1 - \tau_t)u_t\right) + \gamma_{t+1}x \\
&= \quad \Gamma_t u_t - \Gamma_{t+1}\left(\frac{\gamma_{t+1}}{\Gamma_{t+1}}z_t + \frac{\Gamma_t}{\Gamma_{t+1}}u_t\right) + \gamma_{t+1}x \\
&= \quad \gamma_{t+1}(x - z_t).
\end{aligned}
$$

Combining the above inequality with the fact that $\psi_t = L_t d(z_t) + l_t(z_t)$ and with (B.1), we observe:

$$
\begin{aligned}
L_t d(x) &+ l_{t+1}(x) \\
&= \quad L_t d(x) + l_t(x) + \gamma_{t+1}\left(f(x_{t+1}) + \langle \nabla f(x_{t+1}), x - x_{t+1}\rangle\right) \\
&= \quad L_t V_{z_t}(x) + \psi_t + \langle L_t \nabla d(z_t) + s_t, x - z_t\rangle \\
&\quad + \gamma_{t+1}\left(\langle \nabla f(x_{t+1}), x - x_{t+1}\rangle + f(x_{t+1})\right) \\
&\geq \quad L_t V_{z_t}(x) + \psi_t + \gamma_{t+1}\left(f(x_{t+1}) + \langle \nabla f(x_{t+1}), x - x_{t+1}\rangle\right) \\
&\geq \quad L_t V_{z_t}(x) + \Gamma_{t+1}f(x_{t+1}) + \gamma_{t+1}\langle \nabla f(x_{t+1}), x - z_t\rangle + \chi_t.
\end{aligned}
$$

With $\vartheta_t^{(1)} := (L_{t+1} - L_t)\, d(z_{t+1})$, we thus get:

$$
\begin{aligned}
\psi_{t+1} & \\
:= \quad & \min_{x \in Q}\left\{L_{t+1}d(x) + l_{t+1}(x)\right\} \\
= \quad & L_{t+1}d(z_{t+1}) + l_{t+1}(z_{t+1}) \\
= \quad & \vartheta_t^{(1)} + L_t d(z_{t+1}) + l_{t+1}(z_{t+1}) \\
\geq \quad & \vartheta_t^{(1)} + \min_{x \in Q}\left\{L_t d(x) + l_{t+1}(x)\right\} \\
\geq \quad & \vartheta_t^{(1)} + \min_{x \in Q}\left\{L_t V_{z_t}(x) + \Gamma_{t+1}f(x_{t+1}) + \gamma_{t+1}\langle \nabla f(x_{t+1}), x - z_t\rangle + \chi_t\right\}.
\end{aligned}
$$

Let $\vartheta_t^{(2)} := \frac{1}{2}\left(L_t - L_{t+1}\right)\|z_t - \hat{x}_{t+1}\|^2$. Using the construction rule for $\hat{x}_{t+1}$ and (2.13), we obtain:

$$
\begin{aligned}
\psi_{t+1} & \\
\geq \quad & \vartheta_t^{(1)} + L_t V_{z_t}(\hat{x}_{t+1}) + \Gamma_{t+1}f(x_{t+1}) + \gamma_{t+1}\langle \nabla f(x_{t+1}), \hat{x}_{t+1} - z_t\rangle + \chi_t \\
\geq \quad & \vartheta_t^{(1)} + \frac{L_t}{2}\|z_t - \hat{x}_{t+1}\|^2 \\
& + \Gamma_{t+1}f(x_{t+1}) + \gamma_{t+1}\langle \nabla f(x_{t+1}), \hat{x}_{t+1} - z_t\rangle + \chi_t \\
= \quad & \vartheta_t^{(1)} + \vartheta_t^{(2)} + \frac{L_{t+1}}{2}\|z_t - \hat{x}_{t+1}\|^2 \\
& + \Gamma_{t+1}f(x_{t+1}) + \gamma_{t+1}\langle \nabla f(x_{t+1}), \hat{x}_{t+1} - z_t\rangle + \chi_t
\end{aligned}
$$

$$
\begin{aligned}
= \;& \vartheta_t^{(1)} + \vartheta_t^{(2)} + \chi_t \\
& + \Gamma_{t+1}\left(\frac{L_{t+1}}{2\Gamma_{t+1}} \|z_t - \hat{x}_{t+1}\|^2 + f(x_{t+1}) + \tau_t \langle \nabla f(x_{t+1}), \hat{x}_{t+1} - z_t \rangle \right).
\end{aligned}
$$

As $\tau_t^2 \leq \Gamma_{t+1}^{-1}$ and as $x_{t+1} - \tau_t z_t = (1-\tau_t)u_t = u_{t+1} - \tau_t \hat{x}_{t+1}$, this inequality yields to:

$$
\begin{aligned}
\psi_{t+1} \;&\geq\; \vartheta_t^{(1)} + \vartheta_t^{(2)} + \chi_t \\
& + \Gamma_{t+1}\left(\frac{L_{t+1}\tau_t^2}{2} \|z_t - \hat{x}_{t+1}\|^2 + f(x_{t+1}) + \tau_t \langle \nabla f(x_{t+1}), \hat{x}_{t+1} - z_t \rangle \right) \\
&=\; \vartheta_t^{(1)} + \vartheta_t^{(2)} + \chi_t \\
& + \Gamma_{t+1}\left(\frac{L_{t+1}}{2} \|u_{t+1} - x_{t+1}\|^2 + f(x_{t+1}) + \langle \nabla f(x_{t+1}), u_{t+1} - x_{t+1} \rangle \right).
\end{aligned}
$$

It remains to apply (3.10):

$$
\begin{aligned}
\psi_{t+1} \;&\geq\; \vartheta_t^{(1)} + \vartheta_t^{(2)} + \Gamma_{t+1}f(u_{t+1}) + \chi_t \\
&=\; \sum_{k=0}^{t}(L_{k+1} - L_k)\left(d(z_{k+1}) - \frac{1}{2}\|z_k - \hat{x}_{k+1}\|^2 \right) + \Gamma_{t+1}f(u_{t+1}).
\end{aligned}
$$

$\blacksquare$

B.2 Proof of Theorem 7.2

The notation that we subsequently use refers to Algorithm 7.2. We claim:

$$
\epsilon_t := u_t - l_t \leq \left(\frac{5}{6}\right)^t \left(\sum_{j=1}^{m} \bar{y}_j - \mathrm{Tr}(C)/\alpha \right) \qquad \forall\, 0 \leq t \leq T. \tag{B.2}
$$

Proof: We use induction to prove this claim. For $t = 0$, the above statement holds due to the definitions of $l_0 = \mathrm{Tr}(C)/\alpha$ and $u_0 = \sum_{j=1}^{m} \bar{y}_j$.

Assume now that Bound (B.2) is satisfied for a $0 \leq t \leq T - 1$. We show that this bound also holds for $t + 1$. Assume that we call the feasibility oracle $\mathcal{F}$ with input (ϕ_t, δ_t). If the oracle returns a feasible point X_{t+1} with $\langle C, X_{t+1} \rangle_F > \phi_t$, then:

$$
\begin{aligned}
\epsilon_{t+1} \;:=\;& u_{t+1} - l_{t+1} = u_t - \langle C, X_{t+1} \rangle_F < u_t - \phi_t \\
=\;& \frac{1}{2}(u_t - l_t) < \frac{5}{6}(u_t - l_t) \leq \left(\frac{5}{6}\right)^{t+1}(u_0 - l_0),
\end{aligned}
$$

where the last inequality is due to the induction hypothesis.

Otherwise, we have:

$$
\begin{aligned}
\epsilon_{t+1} \;&:=\; u_{t+1} - l_{t+1} = (1 + \delta_t)\phi_t - l_t \\
&=\; \frac{1}{2}\left(u_t + l_t\right) + \delta_t\phi_t - l_t = \frac{1}{2}\left(u_t - l_t\right) + \delta_t\phi_t.
\end{aligned}
$$

As $\delta_t = \frac{2(u_t - l_t)}{3(u_t + l_t)}$, we obtain:

$$
\epsilon_{t+1} = \frac{1}{2}\left(u_t - l_t\right) + \frac{1}{3}\left(u_t - l_t\right) = \frac{5}{6}\left(u_t - l_t\right) \le \left(\frac{5}{6}\right)^{t+1}\left(u_0 - l_0\right),
$$

where the concluding inequality holds by the induction hypothesis.   ∎

B.3 Proof of Proposition 10.1

We use the same notation as in Chapter 10. For an easier reading, we quickly recall the most important expressions. For $V \in \mathcal{S}^n$, we define

$$
g(V) := \frac{\mathcal{A}^*(\exp(V))}{\mathrm{Tr}(\exp(V))} \in \mathbb{R}^m.
$$

Recall that $\zeta_1, \ldots, \zeta_N : \Xi_\zeta \to \mathbb{R}^n$ are independent $\mathcal{N}(0, I_n)$-distributed random vectors and that we write ξ for $(\zeta_1, \ldots, \zeta_N) : \Xi \to \mathbb{R}^{n \times N}$. In addition, we denote by χ_i the random vector $\exp(V/2)\zeta_i$, where $1 \le i \le N$. The random vector $g_\xi(V)$ is defined as:

$$
g_\xi(V) := \frac{\mathcal{A}^*(G_\xi(V))}{\theta_\xi(V)}
$$

with

$$
G_\xi(V) := \frac{\sum_{i=1}^N \chi_i\chi_i^T}{N} \qquad \text{and} \qquad \theta_\xi(V) := \frac{\sum_{i=1}^N \chi_i^T\chi_i}{N}.
$$

We start with the observation:

Assumption B.1 *As the standard multivariate normal distribution $\mathcal{N}(0, I_n)$ is orthogonal invariant, and as both $g(V)$ and $g_\xi(V)$ are invariant under positive scaling, we can assume, without loss of generality, that $\exp(V)$ is diagonal (with positive diagonal entries) and of trace 1. This assumption shall hold for the rest of this section.*

For $i = 1, \ldots, N$, we consider:

$$
D_{\zeta_i} := \chi_i\chi_i^T - \exp(V), \qquad d_{\zeta_i} \;:=\; \mathcal{A}^*(D_{\zeta_i}), \qquad d_\xi := \frac{1}{N}\sum_{i=1}^N d_{\zeta_i},
$$

and:

$$f_{\zeta_i} \;\; := \;\; \chi_i^T \chi_i - 1, \qquad f_\xi := \frac{1}{N} \sum_{i=1}^{N} f_{\zeta_i}.$$

Lemma B.3 *For an appropriately chosen constant $c > 0$, we have for any $i = 1, \ldots, N$:*

a) $\mathbb{E}_{\mathbb{P}_\zeta} [D_{\zeta_i}] = \mathbb{E}_{\mathbb{P}_\zeta} [d_{\zeta_i}] = \mathbb{E}_{\mathbb{P}} [d_\xi] = \mathbb{E}_{\mathbb{P}_\zeta} [f_{\zeta_i}] = \mathbb{E}_{\mathbb{P}} [f_\xi] = 0.$

b) $\mathbb{E}_{\mathbb{P}_\zeta} \left[\exp \left(\|D_{\zeta_i}\|_{(1)} / c \right) \right] \leq \exp(1).$

c) $\mathbb{E}_{\mathbb{P}_\zeta} \left[\exp \left(\frac{\|d_{\zeta_i}\|_\infty}{c\mathcal{L}} \right) \right] \leq \exp(1).$

d) $\mathbb{E}_{\mathbb{P}} \left[\exp \left(\frac{\sqrt{N}\|d_\xi\|_\infty}{c\mathcal{L}\sqrt{\ln(m)}} \right) \right] \leq \exp(1).$

e) $\mathbb{E}_{\mathbb{P}_\zeta} \left[\exp \left(|f_{\zeta_i}| / c \right) \right] \leq \exp(1).$

f) $\mathbb{E}_{\mathbb{P}} \left[\exp \left(\sqrt{N}|f_\xi| / c \right) \right] \leq \exp(1).$

Proof: Assume that the diagonal of $\exp(V)$ is given by the vector $v \in \Delta_n$. Let $1 \leq i \leq n$.

a) We have:

$$\mathbb{E}_{\mathbb{P}_\zeta} [D_{\zeta_i}] = \exp(V/2)\mathbb{E}_{\mathbb{P}_\zeta} \left[\zeta_i \zeta_i^T \right] \exp(V/2) - \exp(V) = 0,$$

where the concluding equality holds as $\zeta_i \sim \mathcal{N}(0, I_n)$. In addition,

$$\mathbb{E}_{\mathbb{P}_\zeta} \left[\zeta_i^T \exp(V)\zeta_i \right] = \sum_{k=1}^{n} v_k = 1,$$

which proves $\mathbb{E}_{\mathbb{P}_\zeta} [f_{\zeta_i}] = 0$. The remaining equations follow immediately.

b) It holds that:

$$\left\| \chi_i \chi_i^T \right\|_{(1)} = \left\| [\exp(V/2)\zeta_i] [\exp(V/2)\zeta_i]^T \right\|_{(1)} = \sum_{k=1}^{n} v_k \zeta_{i,k}^2. \qquad \text{(B.3)}$$

For any $0 < c_1 \leq \frac{1-\exp(-2)}{2} \left(< \frac{1}{2} \right)$, we obtain:

$$\mathbb{E}_{\mathbb{P}_\zeta} \left[\exp \left(c_1 \left\| \chi_i \chi_i^T \right\|_{(1)} \right) \right] = \prod_{k=1}^{n} \mathbb{E}_{\mathbb{P}_\zeta} \left[\exp \left(c_1 v_k \zeta_{i,k}^2 \right) \right]$$

$$= \prod_{k=1}^{n} (1 - 2c_1 v_k)^{-1/2}$$

$$= \exp\left(-\frac{1}{2} \sum_{k=1}^{n} \ln\left(1 - 2c_1 v_k\right)\right)$$

$$\leq \exp\left(-\frac{1}{2} \ln\left(1 - 2c_1\right)\right)$$

$$\leq \exp(1), \tag{B.4}$$

where the first inequality holds as the maximum of $-\sum_{k=1}^{n} \ln\left(1 - 2c_1 v_k\right)$ over the probability simplex is attained at an extreme point (note that the function

$$v \mapsto -\sum_{k=1}^{n} \ln\left(1 - 2c_1 v_k\right)$$

is separable and each of its components is convex). In particular, it holds that:

$$c_1 \mathbb{E}_{\mathbb{P}_\zeta}\left[\left\|\chi_i \chi_i^T\right\|_{(1)}\right] \leq \exp(1).$$

Using Jensen's Inequality, it follows that:

$$\mathbb{E}_{\mathbb{P}_\zeta}\left[\exp\left(\frac{c_1}{2\exp(1)}\left\|\chi_i \chi_i^T - \mathbb{E}_{\mathbb{P}_\zeta}\left[\chi_i \chi_i^T\right]\right\|_{(1)}\right)\right] \leq \exp(1).$$

We conclude by recalling that $\mathbb{E}_{\mathbb{P}_\zeta}\left[\chi_i \chi_i^T\right] = \exp(V)$, that is, it holds that:

$$\chi_i \chi_i^T - \mathbb{E}_{\mathbb{P}_\zeta}\left[\chi_i \chi_i^T\right] = D_{\zeta_i}.$$

c) Let $0 < c_2 \leq \frac{c_1}{2\exp(1)}$. We obtain by b):

$$\mathbb{E}_{\mathbb{P}_\zeta}\left[\exp\left(c_2 \left\|d_{\zeta_i}\right\|_\infty / \mathcal{L}\right)\right] = \mathbb{E}_{\mathbb{P}_\zeta}\left[\exp\left(c_2 \left\|\mathcal{A}^*(D_{\zeta_i})\right\|_\infty / \mathcal{L}\right)\right]$$

$$\leq \mathbb{E}_{\mathbb{P}_\zeta}\left[\exp\left(c_2 \left\|D_{\zeta_i}\right\|_{(1)}\right)\right] \leq \exp(1).$$

d) According to Theorem A.1 (note that we apply Theorem A.1 with $\chi = 1$ and $\sigma_i = \mathcal{L}/c_2$ for any $i = 1, \ldots, N$) and due to Example A.2 (recall that $n \geq 3$), there exists $c_3 > 0$ such that:

$$\mathbb{E}_{\mathbb{P}}\left[\exp\left(\frac{c_2 \sqrt{N} \left\|d_\xi\right\|_\infty}{c_3 \mathcal{L} \sqrt{\ln(m)}}\right)\right] \leq \exp(1).$$

e) Due to (B.3) and (B.4), we obtain:

$$\mathbb{E}_{\mathbb{P}_\zeta}\left[\exp\left(c_1\left|\chi_i^T\chi_i-1\right|/2\right)\right]$$

$$\leq \exp(c_1/2)\left(\mathbb{E}_{\mathbb{P}_\zeta}\left[\exp\left(c_1\sum_{k=1}^{n}v_k\zeta_{i,k}^2\right)\right]\right)^{1/2}$$

$$\leq \exp(c_1/2)\exp(1/2) < \exp(1/4)\exp(1/2) = \exp(3/4),$$

where $0 < c_1 \leq \frac{1-\exp(-2)}{2}$.

f) We observe that the space $(\mathbb{R},\|\cdot\|_1)$ has a regularity constant of 1. Due to Theorem A.1 (we apply Theorem A.1 with $\chi=1$ and $\sigma_i=2/c_1$ for any $i=1,\ldots,N$), there exists a constant $c_4 > 0$ such that:

$$\mathbb{E}_{\mathbb{P}}\left[\exp\left(\frac{c_1\sqrt{N}\,|f_\xi|}{2c_4}\right)\right] = \mathbb{E}_\xi\left[\exp\left(\frac{c_1\left|\sum_{k=1}^{N}f_{\zeta_k}\right|}{2c_4\sqrt{N}}\right)\right] \leq \exp(1).$$

It remains to choose $c \geq \max\{2\exp(1)/c_1, c_3/c_2, 2c_4/c_1\}$. $\blacksquare$

We are ready to prove Proposition 10.1.

Proof of Proposition 10.1: Using the same notation as before, we consider the random element:

$$\beta_\xi := d_\xi - f_\xi g(V).$$

Lemma B.3 implies $\mathbb{E}_{\mathbb{P}}[\beta_\xi] = 0$. In addition, by the same lemma, there exists a constant $c_1 > 0$ such that:

$$\mathbb{E}_{\mathbb{P}}\left[\exp\left(\frac{\sqrt{N}\,\|\beta_\xi\|_\infty}{2c_1\mathcal{L}\sqrt{\ln(m)}}\right)\right]$$

$$\leq \mathbb{E}_{\mathbb{P}}\left[\exp\left(\frac{\sqrt{N}\,\|d_\xi\|_\infty}{2c_1\mathcal{L}\sqrt{\ln(m)}}\right)\exp\left(\frac{\sqrt{N}\,\|f_\xi g(V)\|_\infty}{2c_1\mathcal{L}\sqrt{\ln(m)}}\right)\right]$$

$$\leq \left\{\mathbb{E}_{\mathbb{P}}\left[\exp\left(\frac{\sqrt{N}\,\|d_\xi\|_\infty}{c_1\mathcal{L}\sqrt{\ln(m)}}\right)\right]\right\}^{1/2}\left\{\mathbb{E}_{\mathbb{P}}\left[\exp\left(\frac{\sqrt{N}\,\|f_\xi g(V)\|_\infty}{c_1\mathcal{L}\sqrt{\ln(m)}}\right)\right]\right\}^{1/2}$$

$$\leq \exp(1/2)\left\{\mathbb{E}_{\mathbb{P}}\left[\exp\left(\frac{\sqrt{N}\,|f_\xi|}{c_1\sqrt{\ln(m)}}\right)\right]\right\}^{1/2}$$

$$\leq \exp(1), \tag{B.5}$$

where we use Hölder's Inequality and the facts $\|g(V)\|_\infty \leq \mathcal{L}$ and $\ln(m) > 1$. Let

$$\gamma_\xi := g_\xi(V) - \beta_\xi - g(V).$$

As $\theta_\xi(V) = f_\xi + 1$ and $\mathcal{A}^*(G_\xi(V)) = d_\xi + g(V)$, it holds that:

$$g_\xi(V) = \frac{\mathcal{A}^*(G_\xi(V))}{\theta_\xi(V)} = \frac{d_\xi + g(V)}{f_\xi + 1}.$$

We obtain:

$$\gamma_\xi = \frac{d_\xi + g(V)}{f_\xi + 1} - g(V) - d_\xi + f_\xi g(V) = \frac{f_\xi^2 g(V) - d_\xi f_\xi}{1 + f_\xi}.$$

Consider now the sets:

$$\Pi := \{\xi = (\zeta_1, \ldots, \zeta_N) : |f_\xi| \leq 1/2\} \qquad \text{and} \qquad \hat{\Pi} := \mathbb{R}^{n \times N} \setminus \Pi.$$

When $\xi \in \Pi$, we have:

$$\begin{aligned}
\|\gamma_\xi\|_\infty &= \frac{\left\| f_\xi^2 g(V) - d_\xi f_\xi \right\|_\infty}{|1 + f_\xi|} \leq 2 \left\| f_\xi^2 g(V) - d_\xi f_\xi \right\|_\infty \\
&\leq 2 \left(|f_\xi|^2 \mathcal{L} + |f_\xi| \|d_\xi\|_\infty \right) \leq |f_\xi| \mathcal{L} + \|d_\xi\|_\infty .
\end{aligned} \qquad (B.6)$$

More generally, it holds for any $\xi \in \mathbb{R}^{n \times N}$:

$$\begin{aligned}
\|\gamma_\xi\|_\infty &\leq \|g_\xi(V)\|_\infty + \|\beta_\xi\|_\infty + \|g(V)\|_\infty \\
&\leq \|g_\xi(V)\|_\infty + \|d_\xi\|_\infty + \|f_\xi g(V)\|_\infty + \|g(V)\|_\infty \\
&= \|g_\xi(V)\|_\infty + \|d_\xi\|_\infty + |f_\xi| \|g(V)\|_\infty + \|g(V)\|_\infty \\
&\leq (2 + |f_\xi|) \mathcal{L} + \|d_\xi\|_\infty ,
\end{aligned} \qquad (B.7)$$

which is due to the inequalities $\|g(V)\|_\infty \leq \mathcal{L}$ and $\|g_\xi(V)\|_\infty \leq \mathcal{L}$. Furthermore,

$$\begin{aligned}
\mathbb{P}\left[\hat{\Pi} \right] &= \mathbb{P}\left[|f_\xi| > 1/2 \right] = \mathbb{P}\left[\exp\left(\frac{\sqrt{N}\,|f_\xi|}{c_1} \right) > \exp\left(\frac{\sqrt{N}}{2c_1} \right) \right] \\
&< \exp\left(-\frac{\sqrt{N}}{2c_1} \right) \mathbb{E}_\mathbb{P}\left[\exp\left(\sqrt{N}\,|f_\xi|/c_1 \right) \right] \\
&\leq \exp\left(-\frac{\sqrt{N}}{2c_1} \right) \exp(1),
\end{aligned} \qquad (B.8)$$

where the inequalities follow from Markov's inequality and from statement e) of Lemma B.3, respectively. Choose $c_2 \geq 4c_1$ and observe:

$$\begin{aligned}
&\mathbb{E}_\mathbb{P}\left[\exp\left(\frac{\sqrt{N}\,\|\gamma_\xi\|_\infty}{3c_2 \mathcal{L}\sqrt{\ln(m)}} \right) \right] \\
&= \int_\Pi \exp\left(\frac{\sqrt{N}\,\|\gamma_\xi\|_\infty}{3c_2 \mathcal{L}\sqrt{\ln(m)}} \right) d\mathbb{P} + \int_{\hat{\Pi}} \exp\left(\frac{\sqrt{N}\,\|\gamma_\xi\|_\infty}{3c_2 \mathcal{L}\sqrt{\ln(m)}} \right) d\mathbb{P}.
\end{aligned}$$

By (B.6), Hölder's Inequality, and Lemma B.3, we obtain:

$$\int_{\Pi} \exp\left(\frac{\sqrt{N}\,\|\gamma_\xi\|_\infty}{3c_2\mathcal{L}\sqrt{\ln(m)}}\right) d\mathbb{P}$$
$$\leq \int_{\Pi} \exp\left(\frac{\sqrt{N}\left(|f_\xi|\,\mathcal{L}+\|d_\xi\|_\infty\right)}{3c_2\mathcal{L}\sqrt{\ln(m)}}\right) d\mathbb{P}$$
$$\leq 1\cdot\left\{\mathbb{E}_\mathbb{P}\left[\exp\left(\frac{\sqrt{N}\,|f_\xi|}{c_2\sqrt{\ln(m)}}\right)\right]\right\}^{1/3}\left\{\mathbb{E}_\mathbb{P}\left[\exp\left(\frac{\sqrt{N}\,\|d_\xi\|_\infty}{c_2\mathcal{L}\sqrt{\ln(m)}}\right)\right]\right\}^{1/3}$$
$$\leq \exp\left(1/3\right)\exp\left(1/3\right) = \exp(2/3).$$

Additionally, by (B.7), we have:

$$\int_{\hat{\Pi}} \exp\left(\frac{\sqrt{N}\,\|\gamma_\xi\|_\infty}{3c_2\mathcal{L}\sqrt{\ln(m)}}\right) d\mathbb{P} \leq \int_{\hat{\Pi}} \exp\left(\frac{\sqrt{N}\left(2\mathcal{L}+|f_\xi|\,\mathcal{L}+\|d_\xi\|_\infty\right)}{3c_2\mathcal{L}\sqrt{\ln(m)}}\right) d\mathbb{P}.$$

Hölder's Inequality, the relation $c_2 \geq 4c_1$, Jensen's Inequality, Bound (B.8), and Lemma B.3 imply:

$$\int_{\hat{\Pi}} \exp\left(\frac{\sqrt{N}\left(|f_\xi|\,\mathcal{L}+\|d_\xi\|_\infty\right)}{3c_2\mathcal{L}\sqrt{\ln(m)}}\right) d\mathbb{P}$$
$$\leq \left\{\int_{\hat{\Pi}} 1\,d\mathbb{P}\right\}^{1/3}\left\{\int_{\hat{\Pi}} \exp\left(\frac{\sqrt{N}\,|f_\xi|}{c_2\sqrt{\ln(m)}}\right) d\mathbb{P}\right\}^{1/3}$$
$$\cdot\left\{\int_{\hat{\Pi}} \exp\left(\frac{\sqrt{N}\,\|d_\xi\|_\infty}{c_2\mathcal{L}\sqrt{\ln(m)}}\right) d\mathbb{P}\right\}^{1/3}$$
$$\leq \left\{\mathbb{P}\left[\hat{\Pi}\right]\right\}^{1/3}\left\{\mathbb{E}_\mathbb{P}\left[\exp\left(\frac{\sqrt{N}\,|f_\xi|}{c_2\sqrt{\ln(m)}}\right)\right]\right\}^{1/3}$$
$$\cdot\left\{\mathbb{E}_\mathbb{P}\left[\exp\left(\frac{\sqrt{N}\,\|d_\xi\|_\infty}{c_2\mathcal{L}\sqrt{\ln(m)}}\right)\right]\right\}^{1/3}$$
$$\leq \left\{\mathbb{P}\left[\hat{\Pi}\right]\right\}^{1/3}\left\{\mathbb{E}_\mathbb{P}\left[\exp\left(\frac{\sqrt{N}\,|f_\xi|}{c_1}\right)\right]\right\}^{1/12}$$
$$\cdot\left\{\mathbb{E}_\mathbb{P}\left[\exp\left(\frac{\sqrt{N}\,\|d_\xi\|_\infty}{c_1\mathcal{L}\sqrt{\ln(m)}}\right)\right]\right\}^{1/12}$$
$$\leq \exp(1/3)\exp(1/12)\exp(1/12)\exp\left(-\frac{\sqrt{N}}{6c_1}\right)$$
$$= \exp\left(1/2\right)\exp\left(-\frac{\sqrt{N}}{6c_1}\right).$$

As $c_2 \geq 4c_1$, we obtain:

$$\mathbb{E}_{\mathbb{P}}\left[\exp\left(\frac{\sqrt{N}\,\|\gamma_\xi\|_\infty}{3c_2\mathcal{L}\sqrt{\ln(m)}}\right)\right]$$
$$\leq \exp(2/3) + \exp(1/2)\exp\left(\frac{2\sqrt{N}}{3c_2\sqrt{\ln(m)}} - \frac{\sqrt{N}}{6c_1}\right)$$
$$\leq \exp(2/3) + \exp(1/2)\exp\left(\frac{2\sqrt{N}}{3c_2} - \frac{\sqrt{N}}{6c_1}\right)$$
$$\leq 2\exp(2/3). \tag{B.9}$$

Thus,

$$\mathbb{E}_{\mathbb{P}}\left[\exp\left(\frac{\sqrt{N}\,\|g_\xi(V) - g(V)\|_\infty}{6c_2\mathcal{L}\sqrt{\ln(m)}}\right)\right]$$
$$= \mathbb{E}_{\mathbb{P}}\left[\exp\left(\frac{\sqrt{N}\,\|\beta_\xi + \gamma_\xi\|_\infty}{6c_2\mathcal{L}\sqrt{\ln(m)}}\right)\right]$$
$$\leq \left\{\mathbb{E}_{\mathbb{P}}\left[\exp\left(\frac{\sqrt{N}\,\|\beta_\xi\|_\infty}{3c_2\mathcal{L}\sqrt{\ln(m)}}\right)\right]\right\}^{1/2} \left\{\mathbb{E}_{\mathbb{P}}\left[\exp\left(\frac{\sqrt{N}\,\|\gamma_\xi\|_\infty}{3c_2\mathcal{L}\sqrt{\ln(m)}}\right)\right]\right\}^{1/2}$$
$$\leq \left\{\mathbb{E}_{\mathbb{P}}\left[\exp\left(\frac{\sqrt{N}\,\|\beta_\xi\|_\infty}{2c_1\mathcal{L}\sqrt{\ln(m)}}\right)\right]\right\}^{1/2} \left\{\mathbb{E}_{\mathbb{P}}\left[\exp\left(\frac{\sqrt{N}\,\|\gamma_\xi\|_\infty}{3c_2\mathcal{L}\sqrt{\ln(m)}}\right)\right]\right\}^{1/2}$$
$$\leq \sqrt{2}\exp(1/2)\exp(1/3)$$
$$= \sqrt{2}\exp(5/6), \tag{B.10}$$

where the inequalities are due to Hölder's Inequality, the fact that $3c_2 \geq 2c_1$, Bound (B.5), and Inequality (B.9), respectively.

It remains to find an appropriate bound on the norm of $g(V) - \mathbb{E}_{\mathbb{P}}[g_\xi(V)]$. Recall that $\mathbb{E}_{\mathbb{P}}[\beta_\xi] = 0$. Therefore,

$$\|\mathbb{E}_{\mathbb{P}}[g_\xi(V) - g(V)]\|_\infty = \|\mathbb{E}_{\mathbb{P}}[\gamma_\xi]\|_\infty \leq \mathbb{E}_{\mathbb{P}}\left[\|\gamma_\xi\|_\infty\right]$$
$$= \int_\Pi \|\gamma_\xi\|_\infty\,d\mathbb{P} + \int_{\hat\Pi} \|\gamma_\xi\|_\infty\,d\mathbb{P}$$
$$\leq 2\int_\Pi |f_\xi|^2\,\mathcal{L} + |f_\xi|\,\|d_\xi\|_\infty\,d\mathbb{P} + \int_{\hat\Pi} \|\gamma_\xi\|_\infty\,d\mathbb{P},$$

where the concluding inequality follows from (B.6). As $2\exp(x) \geq x^2$ for any $x \geq 0$, we obtain by Lemma B.3:

$$2\mathcal{L}\int_\Pi |f_\xi|^2\,d\mathbb{P} \leq 2\mathcal{L}\mathbb{E}_{\mathbb{P}}\left[|f_\xi|^2\right]$$
$$= \frac{2c_1^2\mathcal{L}}{N}\mathbb{E}_{\mathbb{P}}\left[\left(\frac{\sqrt{N}\,|f_\xi|}{c_1}\right)^2\right]$$

$$
\begin{aligned}
&\leq \frac{4c_1^2 \mathcal{L}}{N} \mathbb{E}_{\mathbb{P}}\left[\exp\left(\frac{\sqrt{N}\,|f_\xi|}{c_1}\right)\right]\\
&\leq \frac{4\exp(1)c_1^2\mathcal{L}}{N}.
\end{aligned}
$$

Furthermore, the same arguments imply:

$$
\begin{aligned}
2\int_\Pi |f_\xi|\,\|d_\xi\|_\infty\,d\mathbb{P} \;\leq\;& \int_\Pi \sqrt{\ln(m)}\mathcal{L}\,|f_\xi|^2\,d\mathbb{P} + \int_\Pi \frac{\|d_\xi\|_\infty^2}{\sqrt{\ln(m)}\mathcal{L}}\,d\mathbb{P}\\
\leq\;& \mathbb{E}_{\mathbb{P}}\left[\sqrt{\ln(m)}\mathcal{L}\,|f_\xi|^2\right] + \mathbb{E}_{\mathbb{P}}\left[\frac{\|d_\xi\|_\infty^2}{\mathcal{L}\sqrt{\ln(m)}}\right]\\
\leq\;& \frac{2c_1^2\mathcal{L}\sqrt{\ln(m)}}{N}\mathbb{E}_{\mathbb{P}}\left[\exp\left(\frac{\sqrt{N}\,|f_\xi|}{c_1}\right)\right]\\
&+\frac{2c_1^2\mathcal{L}\sqrt{\ln(m)}}{N}\mathbb{E}_{\mathbb{P}}\left[\exp\left(\frac{\sqrt{N}\,\|d_\xi\|_\infty}{c_1\mathcal{L}\sqrt{\ln(m)}}\right)\right]\\
\leq\;& \frac{4\exp(1)c_1^2\mathcal{L}\sqrt{\ln(m)}}{N}.
\end{aligned}
$$

Finally,

$$
\begin{aligned}
&\int_{\hat{\Pi}} \|\gamma_\xi\|_\infty\,d\mathbb{P}\\
\leq\;& \left\{\mathbb{P}\left[\hat{\Pi}\right]\right\}^{1/2}\left\{\mathbb{E}_{\mathbb{P}}\left[\|\gamma_\xi\|_\infty^2\right]\right\}^{1/2}\\
\leq\;& \exp\left(\frac{1}{2}-\frac{\sqrt{N}}{4c_1}\right)\left\{\mathbb{E}_{\mathbb{P}}\left[\|\gamma_\xi\|_\infty^2\right]\right\}^{1/2}\\
\leq\;& \exp\left(\frac{1}{2}-\frac{\sqrt{N}}{4c_1}\right)\left\{\frac{18c_2^2\mathcal{L}^2\ln(m)}{N}\mathbb{E}_{\mathbb{P}}\left[\exp\left(\frac{\sqrt{N}\,\|\gamma_\xi\|_\infty}{3c_2\mathcal{L}\sqrt{\ln(m)}}\right)\right]\right\}^{1/2}\\
\leq\;& \frac{6\exp(5/6)\,c_2\mathcal{L}\sqrt{\ln(m)}}{\sqrt{N}}\exp\left(-\frac{\sqrt{N}}{4c_1}\right)\\
\leq\;& \frac{6\exp(5/6)c_2\mathcal{L}\sqrt{\ln(m)}}{\sqrt{N}}\exp\left(-\frac{\sqrt{N}}{c_2}\right)\\
\leq\;& \frac{6\exp(5/6)c_2^2\mathcal{L}\sqrt{\ln(m)}}{N},
\end{aligned}
$$

where the inequalities hold due to Hölder's Inequality, the fact that $\|\gamma_\xi\|_\infty^2$ is nonnegative for any $\xi \in \mathbb{R}^{n\times N}$, Bound (B.8), the fact that $2\exp(x) \geq x^2$ for any $x \geq 0$, Inequality (B.9), the assumption $c_2 \geq 4c_1$, and the relation $\exp(-x) = [\exp(x)]^{-1} \leq x^{-1}$ for all $x > 0$, respectively. We obtain:

$$\|\mathbb{E}_{\mathbb{P}}\left[g_\xi(V)\right] - g(V)\|_\infty$$

$$\leq \quad \frac{4\exp(1)c_1^2\mathcal{L}}{N} + \frac{4\exp(1)c_1^2\mathcal{L}\sqrt{\ln(m)}}{N} + \frac{6\exp(1)c_2^2\mathcal{L}\sqrt{\ln(m)}}{N}$$

$$\leq \quad \frac{8\exp(1)c_1^2\mathcal{L}\sqrt{\ln(m)}}{N} + \frac{6\exp(1)c_2^2\mathcal{L}\sqrt{\ln(m)}}{N}$$

$$\leq \quad \frac{\exp(1)c_2^2\mathcal{L}\sqrt{\ln(m)}}{2N} + \frac{6\exp(1)c_2^2\mathcal{L}\sqrt{\ln(m)}}{N}$$

$$= \quad \frac{13\exp(1)c_2^2\mathcal{L}\sqrt{\ln(m)}}{2N},$$

which proves the first statement of Proposition 10.1. The above inequality ensures together with Bound (B.10):

$$\mathbb{E}_{\mathbb{P}}\left[\exp\left(\frac{\sqrt{N}\,\|g_\xi(V) - \mathbb{E}_{\mathbb{P}}\left[g_\xi(V)\right]\|_\infty}{6c_2\mathcal{L}\sqrt{\ln(m)}}\right)\right]$$

$$\leq \quad \sqrt{2}\exp(5/6)\exp\left(\frac{13\exp(1)c_2}{12\sqrt{N}}\right)$$

$$\leq \quad \sqrt{2}\exp\left(\frac{5}{6} + \frac{13\exp(1)c_2}{12}\right).$$

Let

$$c_3 := \frac{5}{6} + \frac{\ln(2)}{2} + \frac{13\exp(1)c_2}{12}.$$

By Jensen's Inequality, we observe:

$$\mathbb{E}_{\mathbb{P}}\left[\exp\left(\frac{\sqrt{N}\,\|g_\xi(V) - \mathbb{E}_{\mathbb{P}}\left[g_\xi(V)\right]\|_\infty}{6c_2c_3\mathcal{L}\sqrt{\ln(m)}}\right)\right] \quad \leq \quad \exp(1),$$

which implies, together with the relation $x^2 \leq 2\exp(x)$ for all $x \geq 0$, that

$$\mathbb{E}_{\mathbb{P}}\left[\|g_\xi(V) - \mathbb{E}_{\mathbb{P}}\left[g_\xi(V)\right]\|_\infty^2\right] \leq \frac{72\exp(1)c_2^2c_3^2\mathcal{L}^2\ln(m)}{N}.$$

$\blacksquare$

Bibliography

[AE05]   H. Amann and J. Escher, *Analysis I*, 2nd ed., Birkhäuser, 2005.

[AE08]   ______, *Analysis II*, 2nd ed., Birkhäuser, 2008.

[AHK05]   S. Arora, E. Hazan, and S. Kale, *The multiplicative weights update method: a meta algorithm and applications*, Tech. report, Princeton University, 2005, (available at http://www.cs.princeton.edu/~arora/pubs/MWsurvey.pdf; last accessed on March 13, 2012).

[AK07]   S. Arora and S. Kale, *A combinatorial, primal-dual approach to semidefinite programs*, Proceedings of the 39th Annual ACM Symposium on Theory of Computing, San Diego, California, USA, June 2007 (D. Johnson and U. Feige, eds.), ACM, 2007, pp. 227–236.

[Ali95]   F. Alizadeh, *Interior-Point methods in Semidefinite Programming with applications to Combinatorial Optimization*, SIAM Journal on Optimization **5** (1995), no. 1, 13–51.

[BB09]   M. Baes and M. Bürgisser, *Smoothing Techniques for solving semidefinite programs with many constraints*, Tech. report, ETH Zurich, 2009, (available at http://www.optimization-online.org/DB_FILE/2009/10/2414.pdf; last accessed on March 13, 2012).

[BB10]   ______, *Hedge algorithm and Subgradient methods*, Tech. report, ETH Zurich, 2010, (available at http://www.optimization-online.org/DB_FILE/2009/12/2492.pdf; last accessed on March 13, 2012).

[BB11]   ______, *Hedge algorithm and Dual Averaging schemes*, Tech. report, ETH Zurich, 2011, (available at http://arxiv.org/abs/1112.1275; last accessed on March 13, 2012).

[BBN11]   M. Baes, M. Bürgisser, and A. Nemirovski, *A randomized Mirror-Prox method for solving matrix saddle-point problems*, Tech. report, ETH Zurich / Georgia Institute of Technology, 2011, (available at http://arxiv.org/abs/1112.1274; last accessed on March 13, 2012).

[BGFB94]   S. Boyd, L. El Ghaoui, E. Feron, and V. Balakrishnan, *Linear matrix inequalities in System and Control Theory*, Studies in Applied Mathematics, SIAM, 1994.

[BTN97]   A. Ben-Tal and A. Nemirovski, *Robust truss topology design via Semidefinite Programming*, SIAM Journal on Optimization **7** (1997), no. 4, 991–1016.

[BTN01]   ______, *Lectures on modern Convex Optimization: analysis, algorithms, and engineering applications*, MPS/SIAM Series on Optimization, vol. 2, SIAM, 2001.

[BV04]   S. Boyd and L. Vandenberghe, *Convex Optimization*, Cambridge University Press, 2004.

[CBL06]   Nicolo Cesa-Bianchi and Gabor Lugosi, *Prediction, learning, and games*, Cambridge University Press, 2006.

[CE05]   F. Chudak and V. Eleutério, *Improved approximation schemes for linear programming relaxations of combinatorial optimization problems*, Proceedings of the 11th International Conference on Integer Programming and Combinatorial Optimization, Berlin, Germany, June 2005 (M. Jünger and V. Kaibel, eds.), Lecture Notes in Computer Science, vol. 3509, Springer, 2005, pp. 81–96.

[d'A11]   A. d'Aspremont, *Subsampling algorithms for Semidefinite Programming*, Stochastic Systems **2** (2011), no. 1, 274–305.

[Dan63]   G. Dantzig, *Linear Programming and extensions*, Princeton University Press, 1963.

[dat09]   *Portfolio selection instances from Yahoo Finance*, 2009, (available at http://satt.diegm.uniud.it/index.php?page=portfolio-selection; last accessed on March 13, 2012).

[Ele09]   V. Eleutério, *Finding approximate solutions for large scale linear programs*, Ph.D. thesis, ETH Zurich, 2009.

[Fre99]   R. M. Freund, *Introduction to Semidefinite Programming (SDP)*, Lecture notes, 1999, (available at http://dspace.mit.edu/bitstream/handle/1721.1/35259/15-094 Spring-2002/NR/rdonlyres/Sloan-School-of-Management/15-094Systems-Optimization–Models-and-ComputationSpring2002 /A849A5EB-FBF7-4631-8B52-CBBC667E74EB/0/sdpintro.pdf, last accessed on March 1, 2012).

[FS97]   Y. Freund and R. Schapire, *A decision-theoretic generalization of on-line learning and an application to boosting*, Journal of Computer and System Sciences **55** (1997), no. 1, 119–139.

[GW95]   M. Goemans and D. Williamson, *Improved approximation algorithms for Maximum Cut and Satisfiability problems using Semidefinite Programming*, Journal of the Association for Computing Machinery **42** (1995), no. 6, 1115–1145.

[HJ96]   R. Horn and Ch. Johnson, *Matrix analysis*, Cambridge University Press, 1996.

[HNTW09]   U.-U. Haus, K. Niermann, K. Truemper, and R. Weismantel, *Logic integer programming models for signaling networks*, Journal of Computational Biology **16** (2009), no. 5, 725–743.

[HUL93]   J.-B. Hiriart-Urruty and C. Lemaréchal, *Convex Analysis and minimization algorithms I*, Grundlehren der mathematischen Wissenschaften, vol. 305, Springer, 1993.

[IPS05]   G. Iyengar, D. Phillips, and C. Stein, *Approximation algorithms for semidefinite packing problems with applications to Maxcut and Graph Coloring*, Proceedings of the 11th International Conference on Integer Programming and Combinatorial Optimization, Berlin, Germany, June 2005 (M. Jünger and V. Kaibel, eds.), Lecture Notes in Computer Science, vol. 3509, Springer, 2005, pp. 152–166.

[IPS11]   G. Iyengar, D. J. Phillips, and C. Stein, *Approximating semidefinite packing programs*, SIAM Journal on Optimization **21** (2011), no. 1, 231–268.

[JL84]   W. B. Johnson and J. Lindenstrauss, *Extensions of Lipschitz mappings into a Hilbert space*, Contemp. Math **26** (1984), 189–206.

[JNT08]   A. Juditsky, A. Nemirovski, and C. Tauvel, *Solving variational inequalities with stochastic Mirror-Prox algorithm*, Tech. report, Université J. Fourier Grenoble / Georgia Institute of Technology, 2008, (available at http://arxiv.org/abs/0809.0815; last accessed on March 13, 2012).

[Kar84]   N. Karmarkar, *A new polynomial-time algorithm for Linear Programming*, Combinatorica **4** (1984), 373–395.

[KM72]   V. Klee and G. Minty, *How good is the Simplex algorithm?*, Inequalities 3 (O. Shisha, ed.), Academic Press, 1972, pp. 159–175.

[KW92]   J. Kuczyński and H. Woźniakowski, *Estimating the largest eigenvalue by the Power and Lanczos algorithms with a random start*, SIAM Journal on Matrix Analysis and Applications **13** (1992), no. 4, 1094–1122.

[ML03]   C. Moler and C. Van Loan, *Nineteen dubious ways to compute the exponential of a matrix, twenty-five years later*, SIAM Review **45** (2003), no. 1, 3–49.

[Nem04a]   A. Nemirovski, *Prox-method with rate of convergence $O(1/t)$ for variational inequalities with Lipschitz continuous monotone operators and smooth convex-concave saddle point problems*, SIAM Journal on Optimization **15** (2004), no. 1, 229–251.

[Nem04b]   ______, *Regular Banach spaces and large deviations of random sums*, Tech. report, Georgia Institute of Technology, 2004, (available at http://www2.isye.gatech.edu/~nemirovs/LargeDev2004.pdf; last accessed on March 12, 2012).

[Nes04]   Y. Nesterov, *Introductory lectures on Convex Optimization: a basic course*, Applied Optimization, vol. 87, Kluwer Academic Publishers, 2004.

[Nes05] ———, *Smooth minimization of non-smooth functions*, Mathematical Programming **103** (2005), no. 1, 127–152.

[Nes07] ———, *Smoothing Technique and its applications in Semidefinite Optimization*, Mathematical Programming **110** (2007), no. 2, 245–259.

[Nes09] ———, *Primal-Dual Subgradient methods for convex problems*, Mathematical Programming **120** (2009), no. 1, 221–259.

[NN93] Y. Nesterov and A. Nemirovski, *Interior point polynomial algorithms in Convex Programming*, Studies in Applied Mathematics, no. 13, SIAM, 1993.

[NRT99] A. Nemirovski, C. Roos, and T. Terlaky, *On maximization of quadratic form over intersection of ellipsoids with common center*, Mathematical Programming **86** (1999), no. 3, 463–473.

[NY83] A. Nemirovski and D. Yudin, *Problem complexity and method efficiency in Optimization*, John Wiley, 1983.

[Peñ08] J. Peña, *Nash equilibria computation via Smoothing Techniques*, Optima **78** (2008), 12–13.

[PRP04] N. D. Price, J. L. Reed, and B. O. Palsson, *Genome-scale models of microbial cells: evaluating the consequences of constraints*, Nature Reviews. Microbiology **2** (2004), no. 11, 886–897.

[Ren01] J. Renegar, *A mathematical view of Interior-Point methods in Convex Optimization*, MPS/SIAM Series on Optimization, vol. 1, SIAM, 2001.

[Roc70] R. T. Rockafellar, *Convex Analysis*, Princeton Mathematics Series, vol. 28, Princeton University Press, 1970.

[Roc93] ———, *Lagrange multipliers and optimality*, SIAM Review **35** (1993), no. 2, 183–238.

[Sed] *Webpage of SeDuMi*, (available at http://sedumi.ie.lehigh.edu/; last accessed on March 12, 2012).

[Sho77] N. Shor, *New directions in the development of methods in nonsmooth Optimization*, Kybernetika (1977), no. 6, 78–91, (in Russian).

[Stu99]   J. Sturm, *Using SeDuMi 1.02, a Matlab toolbox for Optimization over symmetric cones*, Optimization Methods and Software **11** (1999), 625–653.

[TRW05]   K. Tsuda, G. Rätsch, and M. K. Warmuth, *Matrix exponentiated gradient updates for on-line learning and Bregman projections*, Journal of Machine Learning Research **6** (2005), 995–1018.

[TS08]   M. Terzer and J. Stelling, *Large-scale computation of elementary flux modes with bit pattern trees*, Bioinformatics **24** (2008), no. 19, 2229–2235.

[Wig58]   E. P. Wigner, *On the distribution of the roots of certain symmetric matrices*, The Annals of Mathematics **67** (1958), no. 2, 325–327.

[WK06]   M. K. Warmuth and D. Kuzmin, *Online variance minimization*, In Proceedings of the 19th Annual Conference on Learning Theory (COLT 06), Springer, 2006, pp. 514–528.

[WKR+07] X. Wu, V. Kumar, Q. J. Ross, J. Ghosh, Q. Yang, H. Motoda, G. J. McLachlan, A. Ng, B. Liu, P. S. Yu, Z. Zhou, M. Steinbach, D. J. Hand, and D. Steinberg, *Top 10 algorithms in Data Mining*, Knowl. Inf. Syst. **14** (2007), no. 1, 1–37.